T0214447

Large-Scale Data Analytics

Aris Gkoulalas-Divanis • Abderrahim Labbi
Editors

Large-Scale Data Analytics

Springer

Editors
Aris Gkoulalas-Divanis
IBM Research – Ireland
Damastown Industrial Estate
Mulhuddart, Ireland

Abderrahim Labbi
IBM Research – Zurich
Rüschlikon, Switzerland

ISBN 978-1-4939-4225-1   ISBN 978-1-4614-9242-9 (eBook)
DOI 10.1007/978-1-4614-9242-9
Springer New York Heidelberg Dordrecht London

Printed on acid-free paper

Springer is part of Springer Science+Business Media (www.springer.com)

Preface

In recent years, we are witnessing a data explosion: almost 90 % of today's data have been produced only in the last 2 years, with data being nowadays produced in the order of Zettabytes! This data comes from various sources, including sensors, social networking sites, mobile phone applications, electornic medical record systems and e-commerce sites, just to name a few. Apart from its massive volume, this data is also characterized by variety (heterogeneity) and velocity (streams of data).

Traditional approaches and algorithms are not able to process and analyze such massive and complex datasets. This has signified the need for a paradigm shift, where new hardware and software technology is emerging to efficiently and reliably manage, store, process, analyze and synthesize very large amounts of complex data generated by massively distributed data sources. Beside their massively distributed nature, which requires new distributed architectures for data analysis, the heterogeneity of such sources imposes significant challenges for the efficient analysis of the data under numerous constraints, such as consistent data integration, data homogenization and scaling, privacy and security preservation. Moreover, the emerging real-world applications in domains such as healthcare, weather forecasting, financial engineering, urban planning, traffic management and environmental monitoring impose extra requirements for large-scale data analysis.

This edited book contains contributions on cutting edge research related to large-scale data analytics in the following core areas: databases, data mining, supercomputing, data visualization and privacy. Our goal is to present to students, researchers, professionals and practitioners the state-of-the-art research, which will help shape up the future of large-scale analytics, leading the way to the design of new approaches and technologies that can analyze and synthesize very large amounts of heterogeneous data, generated by massively distributed data sources.

Each chapter of the book presents a survey of an area in large-scale data analytics, or individual results of the emerging research in the field. Chapters 1 and 2 are devoted to the MapReduce framework. In particular, the first chapter provides a comprehensive survey for a family of approaches and mechanisms of large scale data analysis that have been implemented based on the MapReduce framework. Chapter 2 focuses on optimization approaches for plain MapReduce jobs, as well as for parallel data flow systems. Chapters 3 and 4 present two important application areas of the MapReduce framework: mining tera-scale graphs for patterns and anomalies (Chap. 3), and analyzing customer behavioral data for the Telecom industry (Chap. 4). In Chap. 5, the authors describe a unified heterogeneous architecture that integrates massively threaded shared-memory multiprocessors into MapReduce-based clusters to enable executing Map and Reduce operators on thousands of threads, across multiple GPU devices and nodes. The proposed hybrid system can be used to accelerate machine learning algorithms, such as support vector machines, achieving significant speedup. Chapter 6 is devoted to large-scale social network analysis, offering a comprehensive survey of the state-of-the-art in this area, with focus on parallel algorithms and libraries for the computation of network centrality metrics. An overview of data visualization methods that help users to gain insight into large, heterogeneous, dynamic textual datasets is provided in Chap. 7. The last chapter of the book is devoted to technologies for offering security and privacy at large scale. The authors of this chapter present a novel framework for privacy-preserving, distributed data analysis that is practical for many real-world applications.

We, as editors, are genuinely grateful to all contributors of this book for the time and effort they put into this project, despite the heavy burden that we put on them. We also owe special thanks to the effort of the external reviewers for their help in this effort. Last but not least, we are indepted to Susan Lagerstrom-Fife and Courtney Clark from Springer, for their great support towards the preparation and completion of this work. Their editing suggestions were valuable to improving the organization, readability and appearance of the manuscript.

Mulhuddart, Ireland   Aris Gkoulalas-Divanis
Rüschlikon, Switzerland   Abderrahim Labbi

Contents

Contributors

John Canny Computer Science Division, University of California, Berkeley, CA, USA

Yitao Duan NetEase Youdao, Beijing, China

Christos Faloutsos School of Computer Science, Carnegie Mellon University, Pittsburgh, PA, USA

Michael Granitzer University of Passau, Passau, Germany

Sergio Herrero-Lopez Technologies, Equities and Currency (TEC) Division, SwissQuant Group AG, Zurich, Switzerland

Fabian Hueske Technische Universität Berlin, Berlin, Germany

U Kang Department of Computer Science, KAIST University, Republic of Korea

Wolfgang Kienreich Know-Center Graz, Graz, Austria

David Konopnicki IBM Haifa Research Lab, Haifa, Israel

Mattia Lambertini Department of Computer Science and Engineering, University of Bologna, Bologna, Italy

Elisabeth Lex Know-Center Graz, Graz, Austria

Anna Liu NICTA and University of New South Wales, Sydney, NSW, Australia

Matteo Magnani Department of Information Technology, Uppsala University, 751 05 Uppsala, Sweden

Volker Markl Technische Universität Berlin, Berlin, Germany

Moreno Marzolla Department of Computer Science and Engineering, University of Bologna, Bologna, Italy

Danilo Montesi Department of Computer Science and Engineering, University of Bologna, Bologna, Italy

Carmine Paolino Department of Computer Science, Vrije Universiteit, Amsterdam, The Netherlands

Vedran Sabol Know-Center Graz, Graz, Austria

Sherif Sakr NICTA and University of New South Wales, Sydney, NSW, Australia

Christin Seifert University of Passau, Passau, Germany

Michal Shmueli-Scheuer IBM Haifa Research Lab, Haifa, Israel

John R. Williams Massachusetts Institute of Technology, Cambridge, MA, USA

List of Figures

List of Tables

Acronyms

AHP	Analytical hierarchical process
APA	Austrian Press Agency
API	Application Programmer Interface
APP	Mobile application
AQL	Asterix query language
BFS	Breadth-First Search
BGL	Boost Graph Library
BOW	Bag of Words
C2S	Client to server
CC	Connected components
CDR	Call data record
CMV	Coordinated multiple views
DAG	Directed acyclic Graph
DBMS	Database Management System
DC	Disconnected component
DDL	Data definition language
DFS	Depth-First Search/Distributed File System
DML	Data manipulation language
EC2	Amazon Elastic Compute Cloud
ECC	Elliptic curve cryptography
EDR	Event data/detail record
EFF	Electronic Frontier Foundation
EM	Expectation maximization
FIFO	First In First Out
GCC	Giant connected component
GFS	Google File System
GIM-V	Generalized iterative matrix-vector multiplication
GPU	Graphics processing unit
HDFS	Hadoop Distributed File System
IAAS	Infrastructure As A Service
IE	Information extraction

IRAM	Implicitly Restarted Arnoldi Method
JSON	JavaScript object notation
KD	Knowledge discovery
LAN	Local area network
LINQ	Language INtegrated Query
LSI	Latent semantic indexing
MDS	Multidimensional scaling
MPC	Secure multi-party computation
MPI	Message Passing Interface
MPP	Massively parallel processing/processor
MPPN	Massively parallel processing node
MRF	MapReduce framework
MSA	Massive scale analytics
MST	Minimum spanning tree
MTGL	Multi-Threaded Graph Library
NAS	Network attached storage
NUMA	Non-uniform memory access
ODP	Open Directory Project
PACT	Parallelization contract
PBGL	Parallel Boost Graph Library
PCA	Principal component analysis
POS	Part of speech
PPDM	Privacy preserving data mining
PRG	Pseudo-random number generator
QP	Quadratic programming
RAID	Redundant array of independent/inexpensive disks
RAIN	Redundant array of independent/inexpensive nodes
RDBMS	Relational Database Management System
RDF	Resource Description Framework
RRS	Recursive random search
RWR	Random Walk with Restart
S2S	Server to Server
SAN	Storage area network
SCC	Strongly connected components
SCOPE	Structured Computations Optimized for Parallel Execution
SMO	Sequential minimal optimization
SMP	Symmetric multi-processing machine
SNA	Social network analysis
SNAP	Small-world Network Analysis and Partitioning
SNS	Social network site
SQL	Structured query language
SSSP	Single source shortest path
SVD	Singular value decomposition
SVM	Support vector machine
TCP	Transmission Control Protocol

UDF	User-defined function
UMA	Uniform memory access
URL	Uniform resource locator
VSS	Verifiable secret sharing
ZKP	Zero-knowledge proof

Chapter 1
The Family of Map-Reduce

Sherif Sakr and Anna Liu

Abstract In the last two decades, the continuous increase of computational power
has produced an overwhelming flow of data, which called for a paradigm shift in the
computing architecture and large scale data processing mechanisms. MapReduce
is a simple and powerful programming model that enables easy development of
scalable parallel applications that can process vast amounts of data on large clusters
of commodity machines. MapReduce isolates the application from the details of
running a distributed program, such as issues on data distribution, scheduling and
fault tolerance. However, the original implementation of the MapReduce framework
had some limitations that have been tackled by many research efforts in following
up work. This chapter provides a comprehensive survey for a family of approaches
and mechanisms of large scale data analysis that have been implemented based on
the original father idea of the MapReduce framework, and are currently gaining a
lot of momentum in both research and industrial communities. Some case studies
are discussed as well.

1.1 Introduction

In the last two decades, the continuous increase of computational power has
produced an overwhelming flow of data which called for a paradigm shift in
the computing architecture and large scale data processing mechanisms. Powerful
telescopes in astronomy, particle accelerators in physics, and genome sequencers in
biology are putting massive volumes of data into the hands of scientists. Facebook
collects 15 TB of data each day into a PetaByte-scale data warehouse. Jim Gray,
a database software pioneer and a Microsoft researcher, called the shift a *"fourth
paradigm"* [26]. The first three paradigms were *experimental, theoretical* and,

S. Sakr (✉) • A. Liu
NICTA and University of New South Wales, Sydney, NSW, Australia
e-mail: Sherif.Sakr@nicta.com.au; Anna.Liu@nicta.com.au

A. Gkoulalas-Divanis and A. Labbi (eds.), *Large-Scale Data Analytics*,
DOI 10.1007/978-1-4614-9242-9__1, © Springer Science+Business Media New York 2014

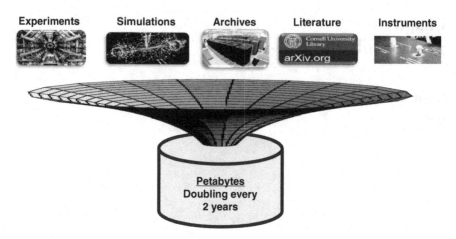

Fig. 1.1 Data explosion in scientific computing [26]

more recently, *computational science*. Gray argued that the only way to cope with this paradigm is to develop a new generation of computing tools to manage, visualize and analyze the data flood. In general, current computer architectures are increasingly imbalanced, where the latency gap between multi-core CPUs and mechanical hard disks is growing every year, which makes the challenges of data-intensive computing much harder to overcome [8]. Hence, there is a crucial need for a systematic and generic approach to tackle these problems with an architecture that can also scale into the foreseeable future. In response, Gray argued that the new trend should instead focus on supporting cheaper clusters of computers to manage and process all this data, instead of focusing on having the biggest and fastest single computer.

Figure 1.1 illustrates an example of the explosion in scientific data which creates major challenges for cutting-edge scientific projects. For example, modern high-energy physics experiments, such as *DZero*,[1] typically generate more than 1 TB of data per day. With datasets growing beyond a few hundreds of terabytes, scientists have no off-the-shelf solutions that they can readily use to manage and analyze these data [26]. Thus, significant human and material resources were allocated to support these data-intensive operations, which led to high storage and management costs.

In general, the growing demand for large-scale data mining and data analysis applications has spurred the development of novel solutions from both the industry (e.g., web-data analysis, click-stream analysis, network-monitoring log analysis) and the sciences (e.g., analysis of data produced by massive-scale simulations, sensor deployments, high-throughput lab equipment) [37]. Although parallel database systems serve some of these data analysis applications, they are expensive, difficult to administer and lack fault-tolerance for long-running queries [34]. MapReduce [16] is a framework which is introduced by Google for programming

---

[1]http://www-d0.fnal.gov/.

commodity computer clusters to perform large-scale data processing in a single pass. The framework is designed in a way that a MapReduce cluster can scale to thousands of nodes in a fault-tolerant manner. An important advantage of this framework is its reliance on a simple and powerful programming model. In addition, MapReduce isolates the application developer from all the complex details of running a distributed program, such as issues on data distribution, scheduling and fault tolerance.

Recently, there has been a great deal of hype about cloud computing [5]. In principle, cloud computing is associated with a new paradigm for the provision of computing infrastructure. This paradigm shifts the location of this infrastructure to the network to reduce the costs associated with the management of hardware and software resources. In particular, cloud computing has promised a number of advantages for hosting the deployments of data-intensive applications, such as:

- Reduced time-to-market by removing or simplifying the time-consuming hardware provisioning, purchasing and deployment processes.
- Reduced monetary cost by following a *pay-as-you-go* business model.
- Unlimited (virtually) throughput by adding servers if the workload increases.

In principle, the success of many enterprises often relies on their ability to analyze expansive volumes of data. In general, cost-effective processing of large datasets had been considered as a nontrivial undertaking. Fortunately, MapReduce frameworks and cloud computing have made it easier than ever for everyone to step into the world of Big data. This technology combination has enabled even small companies to collect and analyze terabytes of data in order to gain a competitive edge. For example, the Amazon Elastic Compute Cloud (EC2)[2] is offered as a commodity that can be purchased and utilised. In addition, Amazon has also provided the Amazon Elastic MapReduce[3] as an online service to easily and cost-effectively process vast amounts of data without the need to worry about time-consuming set-up, management or tuning of computing clusters or the compute capacity upon which they sit. Hence, such services enable third-parties to perform their analytical queries on massive datasets with minimum effort and cost, by abstracting the complexity entailed in building and maintaining computer clusters.

The implementation of the basic MapReduce architecture had some limitations. As a result, many research efforts have been triggered to tackle these limitations by introducing several advancements in the basic architecture in order to improve its performance. This chapter provides a comprehensive survey for a *family* of approaches and mechanisms of large scale data analysis that have been implemented based on the original *father* idea of the MapReduce framework and are currently gaining a lot of momentum in both research and industrial communities. In particular, the remainder of this chapter is organized as follows. Section 1.2 describes the basic architecture of the MapReduce framework. Section 1.3 discusses several techniques that have been proposed to improve the performance and

---

[2]http://aws.amazon.com/ec2/.

[3]http://aws.amazon.com/elasticmapreduce/.

capabilities of the MapReduce framework. Section 1.4 gives an overview of several systems that support high level SQL-like interface for the MapReduce framework, while Sect. 1.5 discusses the hybrid systems that support both MapReduce and SQL-like interfaces. Several case studies are discussed in Sect. 1.6, before we conclude the chapter in Sect. 1.7.

1.2   The MapReduce Framework: Basic Architecture

The MapReduce framework is introduced as a simple and powerful programming model that enables easy development of scalable parallel applications which can process vast amounts of data on large clusters of commodity machines [16, 17]. In particular, the framework is mainly designed to achieve high performance on large clusters of commodity PCs. One of the main advantages of this approach is that it isolates the application from the details of running a distributed program, such as issues on data distribution, scheduling and fault tolerance. In this model, the computation takes a set of input key/value pairs and produces a set of output key/value pairs.

The user of the MapReduce framework expresses the computation using two functions: *Map* and *Reduce*. The Map function takes an input pair and produces a set of intermediate key/value pairs. The MapReduce framework groups together all intermediate values associated with the same intermediate key I and passes them to the Reduce function. The Reduce function receives an intermediate key I with its set of values and merges them together. Typically just zero or one output value is produced per Reduce invocation. The main advantage of this model is that it allows large computations to be easily parallelized and re-executed to be used as the primary mechanism for fault tolerance.

Figure 1.2 illustrates an example MapReduce program expressed in pseudo-code for counting the number of occurrences of each word in a collection of documents. In this example, the map function emits each word plus an associated mark of occurrences, while the reduce function sums together all marks emitted for a particular word. In principle, the design of the MapReduce framework has considered the following main principles [46]:

```
map(String key, String value):          reduce(String key, Iterator values):
// key: document name                    // key: a word
// value: document contents              // values: a list of counts
for each word w in value:                int result = 0;
        EmitIntermediate(w, "1");        for each v in values:
                                                 result += ParseInt(v);
                                         Emit(AsString(result));
```

Fig. 1.2 An example of a MapReduce program [16]

- *Low-Cost Unreliable Commodity Hardware*: Instead of using expensive, high-performance, reliable symmetric multiprocessing (SMP) or massively parallel processing (MPP) machines equipped with high-end network and storage subsystems, the MapReduce framework is designed to run on large clusters of commodity hardware. This hardware is managed and powered by open-source operating systems and utilities so that the cost is kept low.
- *Extremely Scalable RAIN Cluster*: Instead of using centralized RAID-based SAN or NAS storage systems, every MapReduce node has its own local off-the-shelf hard drives. These nodes are loosely coupled in rackable systems that are connected with generic LAN switches. These nodes can be taken out of service with almost no impact to still-running MapReduce jobs. These clusters are called Redundant Array of Independent (and Inexpensive) Nodes (RAIN).
- *Fault-Tolerant yet Easy to Administer*: MapReduce jobs can run on clusters with thousands of nodes or even more. These nodes are not very reliable as at any point in time, a certain percentage of these commodity nodes or hard drives will be out of order. Hence, the MapReduce framework applies straightforward mechanisms to replicate data and launch backup tasks so as to keep still-running processes going. To handle crashed nodes, system administrators simply take crashed hardware off-line. New nodes can be plugged in at any time without much administrative hassle. There is no complicated backup, restore and recovery configurations like the ones that can be seen in many DBMS.
- *Highly Parallel yet Abstracted*: The most important contribution of the Map-Reduce framework is its ability to automatically support the parallelization of task executions. Hence, it allows developers to focus mainly on the problem at hand rather than worrying about the low level implementation details, such as memory management, file allocation, parallel, multi-threaded or network programming. Moreover, MapReduce's shared-nothing architecture [38] makes it much more scalable and ready for parallelization.

Hadoop[4] is an open source Java software that supports data-intensive distributed applications by realizing the implementation of the MapReduce framework. On the implementation level, the Map invocations are distributed across multiple machines by automatically partitioning the input data into a set of M splits. The input splits can be processed in parallel by different machines. Reduce invocations are distributed by partitioning the intermediate key space into R pieces using a partitioning function (e.g. hash(key) mod R). The number of partitions (R) and the partitioning function are specified by the user. Figure 1.3 illustrates an example of the overall flow of a MapReduce operation, which goes through the following sequence of actions:

1. The input files of the MapReduce program are split into M pieces and many copies of the program start up on a cluster of machines.

---

[4]http://hadoop.apache.org/.

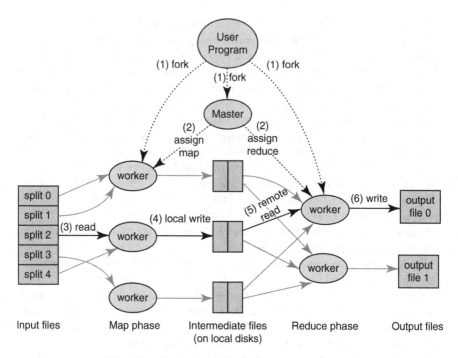

Fig. 1.3 An overview of the flow of execution in a MapReduce operation [16]

2. One of the copies of the program is elected to be the *master* copy, while the rest are considered as *workers* that are assigned their work by the master copy. In particular, there are M map tasks and R reduce tasks to assign. The master picks idle workers and assigns each one a map task or a reduce task.
3. A worker who is assigned a map task reads the contents of the corresponding input split, parses key/value pairs out of the input data and passes each pair to the user-defined Map function. The intermediate key/value pairs produced by the Map function are buffered in memory.
4. Periodically, the buffered pairs are written to local disk, partitioned into R regions by the partitioning function. The locations of these buffered pairs on the local disk are passed back to the master, who is responsible for forwarding these locations to the reduce workers.
5. When a reduce worker is notified by the master about these locations, it reads the buffered data from the local disks of the map workers, which is then sorted by the intermediate keys so that all occurrences of the same key are grouped together. The sorting operation is needed because typically many different keys map to the same reduce task.
6. The reduce worker passes the key and the corresponding set of intermediate values to the user's Reduce function. The output of the Reduce function is appended to a final output file for this reduce partition.

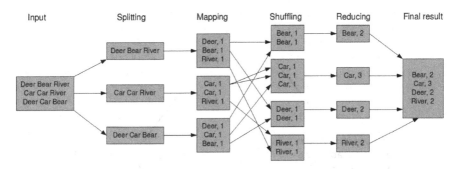

Fig. 1.4 Execution steps of the WordCount example using MapReduce

7. When all map tasks and reduce tasks have been completed, the master program wakes up the user program. At this point, the MapReduce invocation in the user program returns back to the user code.

Figure 1.4 illustrates a sample execution for the example program (WordCount), depicted in Fig. 1.2, using the steps of the MapReduce framework, which are illustrated in Fig. 1.3. During the execution process, the master pings every worker periodically. If no response is received from a worker in a certain amount of time, the master marks the worker as failed. Any map tasks marked completed or in progress by the worker are reset back to their initial idle state and therefore become eligible for scheduling on other workers. Completed map tasks are re-executed on a failure because their output is stored on the local disk(s) of the failed machine and is therefore inaccessible. Completed reduce tasks do not need to be re-executed since their output is stored in a global file system.

1.3   Improvements on the MapReduce Framework

In practice, the basic implementation of MapReduce is very useful for handling data processing and data loading in a heterogenous system with many different storage systems. Moreover, it provides a flexible framework for the execution of complicated functions that can be directly supported in SQL. However, the basic architecture suffers from certain limitations. Dean and Ghemawat [18] reported a set of possible improvements that need to be incorporated into the MapReduce framework. These include:

- MapReduce should take advantage of natural indices whenever possible.
- Most MapReduce output should be left unmerged since there is no benefit of merging them if the next consumer is just another MapReduce program.
- MapReduce users should avoid using inefficient textual formats.

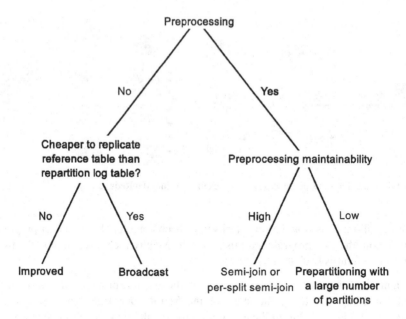

Fig. 1.5 Decision tree for choosing between various join strategies on the MapReduce framework [10]

In the following subsections, we discuss some research efforts that have been conducted in order to deal with these challenges, as well as the different improvements that have been made on the basic implementation of the MapReduce framework in order to achieve these goals.

1.3.1   *Map-Reduce-Merge*

One main limitation of the MapReduce framework is that it does not support the joining of multiple datasets in one task. However, this can still be achieved with additional MapReduce steps. For example, users can map and reduce one dataset and read data from other datasets on the fly. Blanas et al. [10] report on a study that evaluated the performance of different distributed join algorithms (e.g., Repartition Join, Broadcast Join) using the MapReduce framework. Figure 1.5 illustrates a decision tree that summaries the tradeoffs of the considered join strategies, according to the results of that study. Based on statistics, such as the relative data size and the fraction of the join key referenced, this decision tree tries to determine what is the right join strategy for a given circumstance. If data is not preprocessed, the right join strategy depends on the size of the data transferred via the network. If the network cost of broadcasting an input relation R to every node is less expensive than transferring both R and projected L, then the broadcast

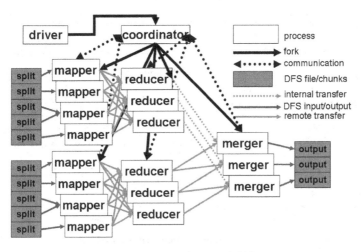

Fig. 1.6 An overview of the Map-Reduce-Merge framework [46]

join algorithm should be used. When preprocessing is allowed, semi-join, per-split semi-join and directed join with enough partitions are the best choices. Semi-join and per-split semi-join offer further flexibility since their preprocessing steps are insensitive to how the log table is organized, and thus suitable for any number of reference tables. In addition, the preprocessing steps of these two algorithms are cheaper since there is no shuffling of the log data.

To tackle the limitation of the join phase in the MapReduce framework, Yang et al. [46] have proposed the Map-Reduce-Merge model that enables the processing of multiple datasets. Figure 1.6 illustrates the framework of this model, where the map phase transforms an input key/value pair $(k1, v1)$ into a list of intermediate key/value pairs $[(k2, v2)]$. The reduce function aggregates the list of values $[v2]$ associated with $k2$ and produces a list of values $[v3]$ which is also associated with $k2$. Note that inputs and outputs of both functions belong to the same lineage (α). Another pair of map and reduce functions produce the intermediate output $(k3, [v4])$ from another lineage (β). Based on keys $k2$ and $k3$, the merge function combines the two reduced outputs from different lineages into a list of key/value outputs $[(k4, v5)]$. This final output becomes a new lineage (γ). If $\alpha = \beta$ then this merge function does a self-merge which is similar to self-join in relational algebra. The main differences between the processing model of this framework and the original MapReduce is the production of a key/value list from the reduce function instead of just that of values. This change is introduced because the merge function needs input datasets organized (partitioned, then either sorted or hashed) by keys and these keys have to be passed into the function to be merged. In the original framework, the reduced output is final. Hence, users pack whatever needed in $[v3]$ while passing $k2$ for the next stage is not required.

Figure 1.7 illustrates a sample execution of the Map-Reduce-Merge framework. In this example, there are two datasets: *Employee* and *Department*, where

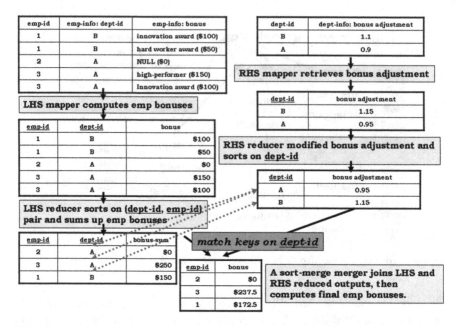

Fig. 1.7 A sample execution of the Map-Reduce-Merge framework [46]

Employee's key attribute is `emp-id` and the Department's key is `dept-id`. The execution of this example query aims to join these two datasets and compute employee bonuses. On the left hand side of Fig. 1.7, a mapper reads Employee entries and computes a bonus for each entry. A reducer then sums up these bonuses for every employee and sorts them by `dept-id`, then `emp-id`. On the right hand side, a mapper reads Department entries and computes bonus adjustments. A reducer then sorts these department entries. At the end, a merger matches the output records from the two reducers on `dept-id` and applies a department-based bonus adjustment on employee bonuses. Yang and Parker [45] have also proposed an approach for improving the Map-Reduce-Merge framework by adding a new primitive, called *traverse*. This primitive can process index file entries recursively, select data partitions based on query conditions and feed only selected partitions to other primitives.

Afrati and Ullman [3] have presented another approach to improve the join phase in the MapReduce framework. This approach begins by identifying the *map-key*, the set of attributes that identify the Reduce process to which a Map process must send a particular tuple. Each attribute of the map-key gets a "*share*", which is the number of buckets into which its values are hashed, to form a component of the identifier of a Reduce process. Relations have their tuples replicated in limited fashion, where the degree of replication depends on the shares for those map-key attributes that are missing from their schema. The approach considers two important special join cases: *chain* joins (represents a sequence of 2-way join operations where the output of one operation in this sequence is used as an input to another operation in a pipelined fashion) and *star* joins (represents joining of a large fact table with several smaller dimension tables). In each case, the proposed algorithm is able to determine the map-key and determine the shares that yield the least replication. The proposed approach is not always superior to the conventional way of using map-reduce to implement joins. However, there are some cases where the proposed approach results in clear wins, such as:

- Analytic queries in which a very large fact table is joined with smaller dimension tables.
- Queries involving paths through graphs with high out-degree, such as the Web or a social network.

1.3.2 MapReduce Online

The basic architecture of the MapReduce framework requires the entire output of each map and reduce task to be *materialized* into a local file before it can be consumed by the next stage. This materialization step allows for the implementation of a simple and elegant checkpoint/restart fault tolerance mechanism. Alvaro et al. [4] proposed a modified architecture in which intermediate data is *pipelined* between operators, while preserving the programming interfaces and fault tolerance models of previous MapReduce frameworks. This pipelining approach provides important advantages to the MapReduce framework, such as:

- The reducers can begin their processing of the data as soon as it is produced by mappers. Therefore, they can generate and refine an approximation of their final answer during the course of execution. In addition, they can provide initial estimates of the results several orders of magnitude faster than the final results.
- It widens the domain of problems to which MapReduce can be applied. For example, it facilitates the ability to design MapReduce jobs that run continuously, accepting new data as it arrives and analyzing it immediately (continuous queries). This allows MapReduce to be used in applications such as event monitoring and stream processing.
- Pipelining delivers data to downstream operators more promptly, which can increase opportunities for parallelism, improve utilization as well as reduce response time.

1.3.3 MRShare

With the emergence of cloud computing, the use of an analytical query processing infrastructure (e.g., Amazon EC2) can be directly mapped to *monetary* value. Taking into account that different MapReduce jobs can perform similar work, there could be many opportunities for sharing the execution of their work. This sharing can reduce the overall amount of work, which consequently leads to the reduction of the monetary charges incurred while utilizing the resources of the processing infrastructure. Nykiel et al. [32] have proposed *MRShare* as a sharing framework which is tailored to transform a batch of queries into a new batch that will be executed more efficiently by merging jobs into groups and evaluating each group as a single query. Based on a defined cost model, they described an optimization problem that aims to derive the optimal grouping of queries in order to avoid performing redundant work and, thus, resulting in significant savings on both processing time and associated cost. In particular, the proposed approach considers exploiting the following sharing opportunities:

- *Sharing Scans.* To share scans between two mapping pipelines M_i and M_j, the input data must be the same. In addition, the key/value pairs should be of the same type. Given that, it becomes possible to merge the two pipelines into a single pipeline and scan the input data only once. However, it should be noted that such combined mapping will produce two streams of output tuples (one for each mapping pipeline M_i and M_j). In order to distinguish the streams at the reducer stage, each tuple is tagged with a `tag()` part. This tagging part is used to indicate the origin mapping pipeline during the reduce phase.
- *Sharing Map Output.* If the map output key and value types are the same for two mapping pipelines M_i and M_j, then the map output streams for M_i and M_j can be shared. In particular, if Map_i and Map_j are applied to each input tuple, then the map output tuples coming only from Map_i are tagged with `tag(i)` only. If a map output tuple was produced from an input tuple by both Map_i and Map_j, it is then tagged by `tag(i)+tag(j)`. Therefore, any overlapping parts of the map output will be shared. In principle, producing a smaller map output leads to savings on sorting and copying intermediate data over the network.
- *Sharing Map Functions.* Sometimes the map functions are identical and thus they can be executed once. At the end of the map stage two streams are produced, each tagged with its job tag. If the map output is shared, then clearly only one stream needs to be generated. Even if only some filters are common in both jobs, it is possible to share parts of map functions.

In practice, sharing scans and sharing map-output yield I/O savings, while sharing map functions (or parts of them) additionally yield CPU savings.

1.3.4 HaLoop

Many data analysis techniques (e.g., the PageRank algorithm, recursive relational queries, social network analysis) require iterative computations. These techniques have a common requirement which is that data are processed iteratively until the computation satisfies a convergence or stopping condition. The basic MapReduce framework does not directly support these iterative data analysis applications. Instead, programmers must implement iterative programs by manually issuing multiple MapReduce jobs and orchestrating their execution using a driver program. In practice, there are two key problems with manually orchestrating an iterative program in MapReduce:

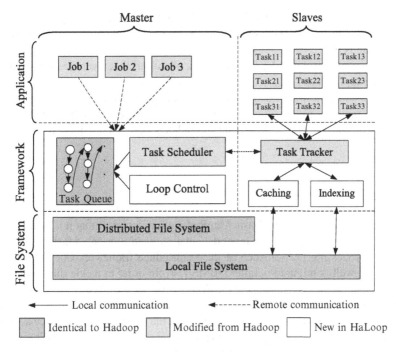

Fig. 1.8   An overview of the HaLoop architecture [11]

- Even though much of the data may be unchanged from iteration to iteration, the data must be re-loaded and re-processed at each iteration, wasting I/O, network bandwidth and CPU resources.
- The termination condition may involve the detection of when a fixpoint has been reached. This condition may itself require an extra MapReduce job on each iteration, again incurring overhead in terms of scheduling extra tasks, reading extra data from disk and moving data across the network.

Bu et al. [11] have presented the *HaLoop* system which is designed to efficiently handle the above types of applications. HaLoop extends the basic MapReduce framework with two main functionalities:

1. A MapReduce cluster can cache the invariant data in the first iteration and then reuse them in later iterations.
2. A MapReduce cluster can cache reducer outputs, which makes checking for a fixpoint more efficient, without an extra MapReduce job.

Figure 1.8 illustrates the architecture of HaLoop as a modified version of the basic MapReduce framework. In order to accommodate the requirements of iterative data analysis applications, HaLoop has incorporated the following changes to the basic Hadoop MapReduce framework:

- It exposes a new application programming interface to users that simplifies the expression of iterative MapReduce programs.

- HaLoop's master node contains a new loop control module that repeatedly starts new map-reduce steps that compose the loop body, until a user-specified stopping condition is met.
- It uses a new task scheduler for iterative applications that leverages data locality in these applications.
- It caches and indices application data on slave nodes. In principle, the task tracker not only manages task execution but also manages caches and indices on the slave node and redirects each task's cache and index accesses to the local file system.

1.3.5 Hadoop++

An important limitation of the Basic MapReduce framework is that it is designed in a way that jobs can only scan the input data in a *sequential*-oriented fashion. Hence, the query processing performance of the MapReduce framework does not match the one of a well-configured parallel DBMS [34]. In order to tackle this challenge, Dittrich et al. [19] have presented the *Hadoop++* system, which aims to boost the query performance of the Hadoop project (the open source implementation of the MapReduce framework) without changing any of the system internals. They achieve this goal by injecting their changes through user-defined functions (UDFs), which only affect the Hadoop system from inside without any external effect. In particular, they introduce the following main changes:

- *Trojan Index*: The original Hadoop implementation does not provide index access due to the lack of a priori knowledge of schema and the MapReduce jobs being executed. Hence, the Hadoop++ system is based on the assumption that if we know the schema and the anticipated MapReduce jobs, then we can create appropriate indices for the Hadoop tasks. In particular, trojan index is an approach to integrate indexing capability into Hadoop in a non-invasive way. These indices are created during the data loading time and thus have no penalty at query time. Each trojan index provides an optional index access path which can be used for selective MapReduce jobs. The scan access path can still be used for other MapReduce jobs. These indices are created by injecting appropriate UDFs inside the Hadoop implementation. Specifically, the main features of trojan indices can be summarized as follows:

  - *No External Library or Engine*: Trojan indices integrate indexing capability natively into the Hadoop framework without imposing a distributed SQL-query engine on top of it.
  - *Non-Invasive*: They do not change the existing Hadoop framework. The index structure is implemented by providing the right UDFs.
  - *Optional Access Path*: They provide an optional index access path which can be used for selective MapReduce jobs. However, the scan access path can still be used for other MapReduce jobs.

- *Seamless Splitting*: Data indexing adds an index overhead for each data split. Therefore, the logical split includes the data as well as the index, as it automatically splits the indexed data at logical split boundaries.
- *Partial Index*: Trojan index need not be built on the entire split. However, it can be built on any contiguous subset of the split as well.
- *Multiple Indexes*: Several trojan indexes can be built on the same split. However, only one of them can be the primary index. During query processing, an appropriate index can be chosen for data access based on the logical query plan and the cost model.

• *Trojan Join*: Similar to the idea of the trojan index, the Hadoop++ system assumes that if we know the schema and the expected workload, then we can co-partition the input data during the loading time. In particular, given any two input relations, they apply the same partitioning function on the join attributes of both the relations at data loading time and place the co-group pairs, having the same join key from the two relations, on the same split and hence on the same node. As a result, join operations can be then processed locally within each node at query time. Implementing the trojan joins does not require any changes to be made to the existing implementation of the Hadoop framework. The only changes are made on the internal management of the data splitting process. In addition, trojan indices can be freely combined with trojan joins.

1.3.6   CoHadoop

In the basic implementation of the Hadoop project, the objective of the data placement policy is to achieve load balancing by distributing the data evenly across the data servers, independently of the intended use of the data. This simple data placement policy works well with most Hadoop applications that access just a *single* file. However, there are other applications that process data from *multiple* files, which can get a significant boost in performance with customized strategies. In these applications, the absence of data co-location increases the data shuffling costs, increases the network overhead and reduces the effectiveness of data partitioning. For example, log processing is a very common usage scenario for the Hadoop framework. In this scenario, data are accumulated in batches from event logs, such as clickstreams, phone call records, application logs or a sequences of transactions. Each batch of data is ingested into Hadoop and stored in one or more HDFS files at regular intervals. Two of the most common operations in log analysis of these applications are (1) joining the log data with some reference data and (2) sessionization, i.e., computing user sessions. The performance of such operations can be significantly improved if they utilize the benefits of data co-location.

CoHadoop [20] is a lightweight extension to Hadoop which is designed to enable co-locating related files at the file system level, while at the same time retaining the good load balancing and fault tolerance properties. CoHadoop introduces a new file property to identify related data files and modify the data placement policy

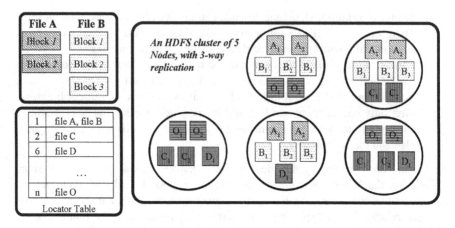

Fig. 1.9 Example file co-location in CoHadoop [20]

of Hadoop to co-locate all copies of those related files in the same server. These changes are designed in a way that retains the benefits of Hadoop, including load balancing and fault tolerance.

In principle, CoHadoop provides a generic mechanism that allows applications to control data placement at the file-system level. In particular, a new file-level property, called a *locator*, is introduced and the Hadoop's data placement policy is modified so that it makes use of this property. Each locator is represented by a unique value (ID), where each file in HDFS is assigned to at most one locator and many files can be assigned to the same locator. Files with the same locator are placed on the same set of datanodes, whereas files with no locator are placed via Hadoop's default strategy. It should be noted that this co-location process involves all data blocks, including replicas. Figure 1.9 shows an example of co-locating two files, *A* and *B*, via a common locator. All of *A*'s two HDFS blocks and *B*'s three blocks are stored on the same set of datanodes. To manage the locator information and keep track of co-located files, CoHadoop introduces a new data structure, the *locator table*, which stores a mapping of locators to the list of files that share this locator. In practice, the CoHadoop extension enables a wide variety of applications to exploit data co-location by simply specifying related files, such as co-locating log files with reference files for joins, co-locating partitions for grouping and aggregation, co-locating index files with their data files and co-locating columns of a table.

1.4 SQL-Like MapReduce Implementations

For programmers, a key appealing feature of the MapReduce framework is that there are only two high-level declarative primitives, *map* and *reduce*, which can be written in any programming language of choice, without worrying about the details of their parallel execution. On the other hand, the MapReduce programming model has its own limitations, such as:

- Its one-input and two-stage data flow is extremely rigid. As we previously discussed, to perform tasks having a different data flow (e.g. joins or n stages), inelegant workarounds have to be devised.
- Custom code has to be written for even the most common operations (e.g. projection and filtering), which leads to the fact that the code is usually difficult to reuse and maintain.
- The opaque nature of the map and reduce functions impedes the ability of the system to perform optimizations.

Moreover, many programmers could be unfamiliar with the MapReduce framework and they would prefer to use SQL (because they are more proficient in) as a high level declarative language to express their task, while leaving all of the execution optimization details to the backend engine. In addition, it is beyond doubt that high level language abstractions enable the underlying system to perform automatic optimization. In what follows, we discuss research efforts to tackle these problems and add the SQL flavor on top of the MapReduce framework.

1.4.1 Pig Latin

Gates et al. [23] have presented a programming language, called *Pig Latin*, which takes a *middle* position between expressing tasks using a high-level declarative querying model in the spirit of SQL, and low-level/procedural programming using MapReduce. Pig Latin is implemented in the scope of the *Apache Pig* project[5] and is used by programmers at Yahoo! for developing data analysis tasks.

Writing a Pig Latin program is similar to specifying a query execution plan (e.g., a data flow graph). To experienced programmers, this method is more appealing than encoding their task as an SQL query and then coercing the system to choose the desired plan through optimizer hints. In general, automatic query optimization has its limits especially with uncataloged data, prevalent user-defined functions and parallel execution, which are all features of the data analysis tasks targeted by the MapReduce framework.

Figure 1.10 shows an example SQL query and its equivalent Pig Latin program. Given a *URL* table with the structure (*url, category, pagerank*), the task of the SQL query is to find each large category and its average pagerank of high-pagerank URLs (>0.2). A Pig Latin program is described as a sequence of steps, where each step represents a single data transformation. This characteristic is appealing to many programmers. At the same time, the transformation steps are described using high-level primitives (e.g. filtering, grouping, aggregation) much like in SQL.

Pig Latin has several other features that are important for casual ad-hoc data analysis tasks. These features include support for a flexible, fully nested data model,

---

[5]http://incubator.apache.org/pig.

SQL	Pig Latin

SELECT category, **AVG**(pagerank)
FROM urls
WHERE pagerank > 0.2
GROUP BY category
HAVING COUNT(*) > 10^6
	good_urls = **FILTER** urls **BY** pagerank > 0.2;
	groups = **GROUP** good_urls **BY** category;
	big_groups = **FILTER** groups **BY** COUNT(good_urls)>10^6;
	output = **FOREACH** big_groups **GENERATE**
	category, **AVG**(good_urls.pagerank);

Fig. 1.10 An example SQL query and its equivalent Pig Latin program [23]

extensive support for user-defined functions and the ability to operate over plain input files without any schema information. In particular, Pig Latin has a simple data model consisting of the following four types:

- *Atom*: An atom contains a simple atomic value, such as a string or a number, e.g., "alice".
- *Tuple*: A tuple is a sequence of fields, each of which can be any of the data types, e.g., ("alice", "lakers").
- *Bag*: A bag is a collection of tuples with possible duplicates. The schema of the constituent tuples is flexible, where not all tuples in a bag need to have the same number and type of fields

  e.g., $\left\{ \begin{array}{l} \text{("alice", "lakers")} \\ \text{("alice", ("iPod", "apple"))} \end{array} \right\}$

- *Map*: A map is a collection of data items, where each item has an associated key through which it can be looked up. As with bags, the schema of the constituent data items is flexible. However, the keys are required to be data atoms, e.g.,

  $\left\{ \begin{array}{l} \text{"k1"} \rightarrow \text{("alice", "lakers")} \\ \text{"k2"} \rightarrow \text{"20"} \end{array} \right\}$

To accommodate specialized data processing tasks, Pig Latin has extensive support for user-defined functions. The input and output of UDFs in Pig Latin follow its fully nested data model. Pig Latin is architected such that the parsing of the Pig Latin program and the logical plan construction is independent of the execution platform. Only the compilation of the logical plan into a physical plan depends on the specific execution platform chosen. Currently, Pig Latin programs are compiled into sequences of MapReduce jobs, which are executed using the Hadoop MapReduce environment.

In particular, a Pig Latin program goes through a series of transformation steps [33] before being executed, as depicted in Fig. 1.11. The parsing steps verifies that the program is syntactically correct and that all referenced variables are defined. The output of the parser is a canonical logical plan with a one-to-one correspondence between Pig Latin statements and logical operators, which are arranged in a *directed acyclic graph* (DAG). The logical plan generated by the parser is passed through a logical optimizer. In this stage, logical optimizations, such as projection pushdown, are carried out. The optimized logical plan is then

Fig. 1.11 Pig compilation
and execution steps [33]

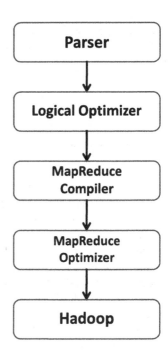

compiled into a series of MapReduce jobs, which are then passed through another optimization phase. The DAG of optimized MapReduce jobs is then topologically sorted and jobs are submitted to Hadoop for execution.

1.4.2   Sawzall

Sawzall [35] is a scripting language used at Google on top of MapReduce. A Sawzall program defines the operations to be performed on a single record of the data. There is nothing in the language to enable examining multiple input records simultaneously, or even to have the contents of one input record influence the processing of another. The only output primitive in the language is the *emit* statement, which sends data to an external aggregator (e.g., Sum, Average, Maximum, Minimum) that gathers the results from each record, after which the results are correlated and processed. The authors argue that aggregation is done outside the language for a couple of reasons: (1) a more traditional language can use the language to correlate results but some of the aggregation algorithms are sophisticated and are best implemented in a native language and packaged in some form, and (2) drawing an explicit line between filtering and aggregation enables a high degree of parallelism and hides the parallelism from the language itself.

Figure 1.12 depicts an example Sawzall program where the first three lines declare the aggregators *count, total* and *sum of squares*. The keyword *table*

```
count: table sum of int;
total: table sum of float;
sumOfSquares: table sum of float;
x: float = input;
emit count $<$- 1;
emit total $<$ -x;
emit sumOfSquares $<$- x * x;
```

Fig. 1.12 An example of a Sawzall program [35]

introduces an aggregator type, which are called tables in Sawzall even though they may be singletons. These particular tables are *sum* tables which add up the values emitted to them, *ints* or *floats* as appropriate. The Sawzall language is implemented as a conventional compiler, written in C++, whose target language is an interpreted instruction set, or byte-code. The compiler and the byte-code interpreter are part of the same binary, so the user presents source code to Sawzall and the system executes it directly. It is structured as a library with an external interface that accepts source code which is then compiled and executed, along with bindings to connect to externally-provided aggregators. The datasets of Sawzall programs are often stored in *Google File System* (GFS) [24]. The business of scheduling a job to execute on a cluster of machines is handled by software, called *Workqueue*, which creates a large-scale time sharing system out of an array of computers and their disks. It schedules jobs, allocates resources, reports status and collects the results.

1.4.3 SQL/MapReduce

In general, a user-defined function is a powerful database feature that allows users to customize database functionality. Friedman et al. [22] introduced the SQL/MapReduce (SQL/MR) UDF framework, which is designed to facilitate parallel computation of procedural functions across hundreds of servers working together as a single relational database. The framework is implemented as part of the *Aster Data Systems*[6] nCluster shared-nothing relational database.

The framework leverages ideas from the MapReduce programming paradigm to provide users with a straightforward API through which they can implement a UDF in the language of their choice. Moreover, it allows maximum flexibility as the output schema of the UDF is specified by the function itself at query plan-time. This means that a SQL/MR function is polymorphic as it can process arbitrary input because its behavior, as well as output schema, are dynamically determined by information available at query plan-time. This also increases reusability as the same SQL/MR function can be used on inputs with many different schemas or

---

[6]http://www.asterdata.com/.

Fig. 1.13 Basic syntax of
SQL/MR query function [22]

```
SELECT ...
FROM functionname(
    ON table-or-query
    [PARTITION BY expr, ...]
    ORDER BY expr, ...]
    [clausename(arg, ...) ...]
    )
```

with different user-specified parameters. In particular, SQL/MR allows the user to write custom-defined functions in any programming language and insert them into queries that otherwise leverage traditional SQL functionality. A SQL/MR function is defined in a manner that is similar to MapReduce's map and reduce functions.

The syntax for using a SQL/MR function is depicted in Fig. 1.13, where the SQL/MR function invocation appears in the SQL *FROM* clause and consists of the function name followed by a set of clauses that are enclosed in parentheses. The *ON* clause specifies the input to the invocation of the SQL/MR function. It is important to note that the input schema to the SQL/MR function is specified implicitly at query plan-time in the form of the output schema for the query used in the ON clause.

In practice, a SQL/MR function can be either a mapper (*Row* function) or a reducer (*Partition* function). The definitions of row and partition functions ensure that they can be executed in parallel in a scalable manner. In the *Row* function, each row from the input table or query will be operated on by exactly one instance of the SQL/MR function. Semantically, each row is processed independently, allowing the execution engine to control parallelism. For each input row, the row function may emit zero or more rows. In the *Partition* function, each group of rows, as defined by the *PARTITION BY* clause, will be operated on by exactly one instance of the SQL/MR function. If the *ORDER BY* clause is provided, the rows within each partition are provided to the function instance in the specified sort order. Semantically, each partition is processed independently, allowing parallelization by the execution engine at the level of a partition. For each input partition, the SQL/MR partition function may output zero or more rows.

1.4.4   SCOPE

SCOPE (Structured Computations Optimized for Parallel Execution) is a scripting language which is targeted for large-scale data analysis and is used for a variety of data analysis and data mining applications inside Microsoft [13]. SCOPE is a declarative language. It allows users to focus on the data transformations required to solve the problem at hand and hides the complexity of the underlying platform and implementation details. The SCOPE compiler and optimizer are responsible for generating an efficient execution plan and the runtime for executing the plan with minimal overhead.

SQL-Like	MapReduce-Like

SELECT query,
          COUNT(*) AS count
FROM "search.log"
USING LogExtractor
GROUP BY query
HAVING count > 1000
ORDER BY count DESC;
OUTPUT TO "qcount.result";
	e = EXTRACT query
	    FROM "search.log" USING LogExtractor;
	s1 = SELECT query, COUNT(*) as count FROM e
	    GROUP BY query;
	s2 = SELECT query, count FROM s1
	    WHERE count > 1000;
	s3 = SELECT query, count FROM s2
	    ORDER BY count DESC;
	OUTPUT s3 TO "qcount.result";

Fig. 1.14 Two equivalent SCOPE scripts in SQL-like style and in MapReduce-like style [13]

Like SQL, data is modeled as sets of rows composed of typed columns. SCOPE is highly extensible. Users can easily define their own functions and implement their own versions of operators: extractors (parsing and constructing rows from a file), processors (row-wise processing), reducers (group-wise processing) and combiners (combining rows from two inputs). This flexibility greatly extends the scope of the language and allows users to solve problems that cannot be easily expressed in traditional SQL. SCOPE provides a functionality which is similar to that of SQL views. This feature enhances modularity and code reusability. It is also used to restrict access to sensitive data. SCOPE supports writing a program using traditional SQL expressions or as a series of simple data transformations.

Figure 1.14 illustrates two equivalent scripts in two different styles that are used to find from a search log queries that have been requested at least 1,000 times. In the MapReduce-like style, the *EXTRACT* command extracts all query string from the log file. The first *SELECT* command counts the number of occurrences of each query string. The second *SELECT* command retains only rows with a count greater than 1,000. The third *SELECT* command sorts the rows on count. Finally, the *OUTPUT* command writes the result to a file.

Microsoft has developed a distributed computing platform, called *Cosmos*, for storing and analyzing massive data sets. Cosmos is designed to run on large clusters consisting of thousands of commodity servers. Figure 1.15 shows the main components of the Cosmos platform, described as follows:

- *Cosmos Storage*: A distributed storage subsystem designed to reliably and efficiently store extremely large sequential files.
- *Cosmos Execution Environment*: An environment for deploying, executing and debugging distributed applications.
- *SCOPE*: A high-level scripting language for writing data analysis jobs. The SCOPE compiler and optimizer translate these scripts to efficient parallel execution plans.

The Cosmos Storage System is an append-only file system that reliably stores petabytes of data. The system is optimized for large sequential I/O. All writes are append-only and concurrent writers are serialized by the system. Data is distributed

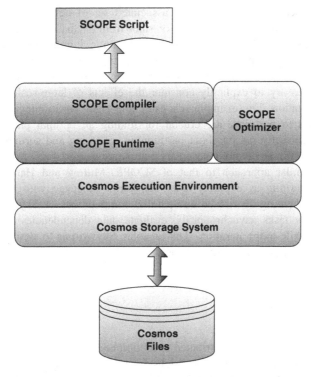

Fig. 1.15 The SCOPE/Cosmos execution platform [13]

and replicated for fault tolerance and compressed to save storage and increase I/O throughput. In Cosmos, an application is modeled as a dataflow graph: a directed acyclic graph with vertices representing processes and edges representing data flows. The runtime component of the execution engine is called the *Job Manager*, which represents the central and coordinating process for all processing vertices within an application.

The SCOPE scripting language resembles SQL but with C# expressions. Thus, it reduces the learning curve for users and eases the porting of existing SQL scripts into SCOPE. Moreover, SCOPE expressions can use C# libraries, where custom C# classes can compute functions of scalar values, or manipulate whole rowsets. A SCOPE script consists of a sequence of commands which are data transformation operators that take one or more rowsets as input, perform some operation on the data and output a rowset. Every rowset has a well-defined schema to which all its rows must adhere. The SCOPE compiler parses the script, checks the syntax and resolves names. The result of the compilation is an internal parse tree which is then translated to a physical execution plan. A physical execution plan is a specification of a Cosmos job, which describes a data flow DAG where each vertex is a program and each edge represents a data channel. The translation into an execution plan is performed by traversing the parse tree in a bottom-up manner.

For each operator, SCOPE has an associated set of default implementation rules. Many of the traditional optimization rules from database systems are clearly also applicable in this new context, for example, removing unnecessary columns, pushing down selection predicates and pre-aggregating when possible. However, the highly distributed execution environment offers new opportunities and challenges, making it necessary to explicitly consider the effects of large-scale parallelism during optimization. For example, choosing the right partitioning scheme and deciding when to partition, are crucial for finding an optimal plan. It is also important to correctly reason about partitioning, grouping and sorting properties and their interaction, to avoid unnecessary computations [49].

Using a similar approach to that of SCOPE, Murray and Hand [31] have presented *Skywriting* as a purely-functional script language with its execution engine for performing distributed and parallel computations. A Skywriting script can create new tasks asynchronously, evaluate data dependencies and perform unbounded (while-loop) iteration. This enables Skywriting to describe a more general class of distributed computations.

1.4.5 DryadLINQ

Dryad is a general-purpose distributed execution engine introduced by Microsoft for coarse-grain data-parallel applications [27]. A Dryad application combines computational *vertices* with communication *channels* to form a dataflow graph. Dryad runs the application by executing the vertices of this graph on a set of available computers, communicating as appropriate through files, TCP pipes and shared-memory FIFOs. The Dryad system offers to the developer fine control over the communication graph, as well as the subroutines that live at its vertices. A Dryad application developer can specify an arbitrary directed acyclic graph to describe the application's communication patterns and express the data transport mechanisms (files, TCP pipes and shared-memory FIFOs) between the computation vertices. This direct specification of the graph gives the developer greater flexibility to easily compose basic common operations, leading to a distributed analogue of *piping* together traditional Unix utilities, such as grep, sort and head.

Dryad is notable for allowing graph vertices (and computations in general) to use an arbitrary number of inputs and outputs, while MapReduce restricts all computations to take a single input set and generate a single output set. The overall structure of a Dryad job is determined by its communication flow. A job is a directed acyclic graph where each vertex is a program and edges represent data channels. It is a logical computation graph that is automatically mapped onto physical resources by the runtime. At run time, each channel is used to transport a finite sequence of structured items. A Dryad job is coordinated by a process called the *Job Manager* that runs either within the cluster or on a user's workstation with network access to the cluster. The job manager contains the application-specific code to construct the job's communication graph along with library code to schedule the work across the available resources. All data is sent directly between vertices and thus the job manager is only responsible for control decisions and is not a bottleneck for any data transfers. Therefore, much of the simplicity of the Dryad scheduler and fault-tolerance model come from the assumption that vertices are deterministic.

Dryad has its own high-level language called DryadLINQ [47]. It generalizes execution environments, such as SQL and MapReduce, in two ways: (1) adopting an expressive data model of strongly typed .NET objects and (2) supporting general-purpose imperative and declarative operations on datasets within a traditional high-level programming language. DryadLINQ[7] exploits LINQ (Language INtegrated Query,[8] a set of .NET constructs for programming with datasets) to provide a powerful hybrid of declarative and imperative programming. The system is designed to provide flexible and efficient distributed computation in any LINQ-enabled programming language including C#, VB and F#.[9] Objects in DryadLINQ datasets can be of any .NET type, making it easy to compute with data such as image patches, vectors and matrices. In practice, a DryadLINQ program is a sequential program composed of LINQ expressions that perform arbitrary side-effect-free transformations on datasets and can be written and debugged using standard .NET development tools. The DryadLINQ system automatically translates the data-parallel portions of the program into a distributed execution plan which is then passed to the Dryad execution platform. Figure 1.16 illustrates the flow of execution when a program is executed by DryadLINQ [47].

1. When a .NET user application runs, it creates a DryadLINQ expression object.
2. The application triggers a data-parallel execution, where the expression object is handed to DryadLINQ.
3. DryadLINQ compiles the LINQ expression into a distributed Dryad execution plan. In particular, it performs the following tasks:

   (a) Decomposes the expression into subexpressions, where each expression can be assigned to run in a separate Dryad vertex.
   (b) Generates the code and static data for the remote Dryad vertices.
   (c) Generates the serialization code for the required data types.

4. DryadLINQ invokes a custom Dryad job manager.
5. The job manager creates the job graph and schedules the vertices as resources become available.
6. Each Dryad vertex executes a vertex-specific program as created in Step 3(b).
7. When the Dryad job completes successfully, it writes the data to the output table(s).

---

[7]http://research.microsoft.com/en-us/projects/dryadlinq/.

[8]http://msdn.microsoft.com/en-us/netframework/aa904594.aspx.

[9]http://research.microsoft.com/en-us/um/cambridge/projects/fsharp/.

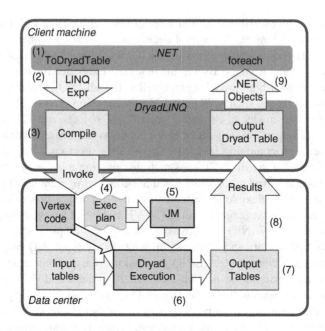

Fig. 1.16 LINQ-expression execution in DryadLINQ [47]

8. The job manager process terminates and returns control back to DryadLINQ, which creates objects encapsulating the outputs of the execution. These objects may be used as inputs to subsequent expressions in the user program.
9. Control returns to the user application. The iterator interface over a DryadTable allows the user to read its contents as .NET objects.
10. The application may generate subsequent DryadLINQ expressions that can be executed by a repetition of Steps 2–9.

1.4.6 Jaql

Jaql[10] is a query language which is designed for Javascript Object Notation (JSON),[11] a data format that has become popular because of its simplicity and modeling flexibility. JSON is a simple, yet flexible way to represent data that ranges from flat, relational data to semi-structured, XML data. Jaql is primarily used to analyze large-scale semi-structured data. It is a functional, declarative query language which rewrites high-level queries (when appropriate) into a low-level

---

[10]http://code.google.com/p/jaql/.

[11]http://www.json.org/.

```
import myrecord;
count Fields = fn(records) (
  records
  -> transform myrecord: :names($)
  -> expand
  -> group by fName = $ as occurrences
     into { name: fName, num: count (occurrences) }
);
read(hdfs("docs.dat"))
-> countFields()
-> write(hdfs("fields.dat"));
```

Fig. 1.17 A sample Jaql script [9]

query, consisting of Map-Reduce jobs that are evaluated using the Apache Hadoop project. Core features include user extensibility and parallelism. Jaql consists of a scripting language and compiler, as well as a runtime component [9]. It is able to process data with no schema or only a partial schema. However, Jaql can also exploit rigid schema information when it is available, for both type checking and improved performance.

Jaql uses a very simple data model; a *JDM value* is either an atom, an array or a record. Most common atomic types are supported by Jaql, including strings, numbers, nulls and dates. Arrays and records are compound types that can be arbitrarily nested. In more detail, an array is an ordered collection of values and can be used to model data structures, such as vectors, lists, sets or bags. A record is an unordered collection of name-value pairs and can model structs, dictionaries, and maps. Despite its simplicity, JDM is very flexible. It allows Jaql to operate with a variety of different data representations for both input and output, including delimited text files, JSON files, binary files, Hadoop's SequenceFiles, relational databases, key-value stores or XML documents. Functions are first-class values in Jaql. They can be assigned to a variable and are high-order in that they can be passed as parameters or used as a return value. Functions are the key ingredient for reusability as any Jaql expression can be encapsulated in a function, and a function can be parameterized in powerful ways.

Figure 1.17 depicts an example of a Jaql script that consists of a sequence of operators. The read operator loads raw data, in this case from Hadoop's Distributed File System (HDFS), and converts it into Jaql values. These values are subsequently processed by the countFields subflow, which extracts field names and computes their frequencies. Finally, the write operator stores the result back into HDFS. In general, the core expressions of the Jaql scripting language include:

1. *Transform*: The transform expression applies a function (or projection) to every element of an array to produce a new array. It has the form e1->transform e2, where e1 is an expression that describes the input array and e2 is applied to each element of e1.

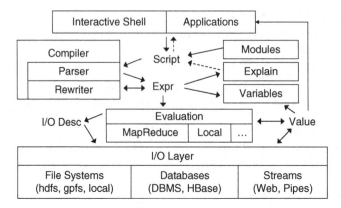

Fig. 1.18 The Jaql system architecture [9]

2. *Expand*: The expand expression is most often used to unnest an input array. It differs from transform in two primary ways: (1) e2 must produce a value v that is an array type, and (2) each of the elements of v is returned to the output array, thereby removing one level of nesting.
3. *Group by*: Similar to SQL's GROUP BY, Jaql's group-by expression partitions its input on a grouping expression and applies an aggregation expression to each group.
4. *Filter*: The filter expression, e− >filter p, retains input values from e for which predicate p evaluates to true.
5. *Join*: The join expression supports equijoin of 2 or more inputs. All of the options for inner and outer joins are also supported.
6. *Union*: The union expression is a Jaql function that merges multiple input arrays into a single output array. It has the form: union(e_1, . . .) where each e_i is an array.
7. *Control-Flow*: The two most commonly used control-flow expressions in Jaql are if-then-else and block expressions. The if-then-else expression is similar to conditional expressions found in most scripting and programming languages. A block establishes a local scope where zero or more local variables can be declared and the last statement provides the return value of the block.

At a high-level, the Jaql system architecture, depicted in Fig. 1.18, is similar to most database systems. Scripts are passed into the system from the interpreter or an application, compiled by the parser and rewrite engine, and either explained or evaluated over data from the I/O layer. The storage layer is similar to a federated database. It provides an API to access data of different systems, including local or distributed file systems (e.g., Hadoop's HDFS), database systems (e.g., DB2, Netezza, HBase), or from streamed sources like the Web. Unlike federated databases, however, most of the accessed data is stored within the same cluster and the I/O API describes data partitioning, which enables parallelism with data affinity during evaluation. Jaql derives much of this flexibility from Hadoop's I/O

API. It reads and writes many common file formats (e.g., delimited files, JSON text, Hadoop Sequence files). Custom adapters are easily written to map a data set to or from Jaql's data model. The input can even simply be values constructed in the script itself. The Jaql interpreter evaluates the script locally on the computer that compiled the script, but spawns interpreters on remote nodes using MapReduce. The Jaql compiler automatically detects parallelization opportunities in a Jaql script and translates it to a set of MapReduce jobs.

1.5   Hybrid Systems

Originally, the applications of the MapReduce framework have been mainly focusing on analyzing very large non-structured datasets, e.g., web indexing, text analytics, and graph data mining. Recently, however, as MapReduce is steadily developing into the de facto data analysis standard, it repeatedly becomes employed for querying structured data [7]. For a long time, relational database and its standard query language (i.e., SQL) has dominated the deployments of data warehousing systems and data analysis on structured data. Therefore, there has been an increasing interest in combining MapReduce and traditional database systems in an effort to maintain the benefits of both worlds. In the following section, we present some systems that have been designed to achieve this goal of integrating the two environments.

1.5.1   Hive

The *Hive* project[12] is an open-source data warehousing solution which has been built by the Facebook Data Infrastructure Team on top of the Hadoop environment [40]. The main goal of this project is to bring the familiar relational database concepts (e.g., tables, columns, partitions) and a subset of SQL to the unstructured world of Hadoop, while still maintaining the extensibility and flexibility that Hadoop enjoyed. Thus, it supports all the major primitive types (e.g., integers, floats, strings) as well as complex types (e.g., maps, lists, structs).

Hive supports queries expressed in an SQL-like declarative language, called *HiveQL*,[13] and therefore can be easily understood by anyone who is familiar with SQL. These queries are compiled into MapReduce jobs that are executed using Hadoop. In addition, HiveQL enables users to plug in custom MapReduce scripts into queries. For example, the canonical MapReduce word count example on a table of documents (Fig. 1.2) can be expressed in HiveQL as depicted in

---

[12]http://hadoop.apache.org/hive/.

[13]http://wiki.apache.org/hadoop/Hive/LanguageManual.

```
FROM (
    MAP doctext USING 'python wc_mapper.py' AS (word, cnt)
    FROM docs
    CLUSTER BY word
) a
REDUCE word, cnt USING 'python wc_reduce.py';
```

Fig. 1.19 An example HiveQl query [40]

Fig. 1.19, where the *MAP* clause indicates how the input columns (*doctext*) can be transformed using a user program ('python wc_mapper.py') into output columns (*word* and *cnt*). The *REDUCE* clause specifies the user program to invoke ('python wc_reduce.py') on the output columns of the subquery.

HiveQL supports *Data Definition Language* (DDL) statements, which can be used to create, drop and alter tables in a database [41]. It allows users to load data from external sources and insert query results into Hive tables, via the load and insert *Data Manipulation Language* (DML) statements, respectively. However, HiveQL currently does not support the update and deletion of rows in existing tables (in particular, INSERT INTO, UPDATE and DELETE statements), which allows the use of very simple mechanisms to deal with concurrent read and write operations without implementing complex locking protocols. The metastore component is the Hive's system catalog which stores metadata about the underlying table. This metadata is specified during table creation and reused every time the table is referenced in HiveQL. The metastore distinguishes Hive as a traditional warehousing solution when compared with similar data processing systems that are built on top of MapReduce-like architectures, such as Pig Latin [33].

1.5.2 HadoopDB

Parallel database systems have been commercially available for nearly two decades and there are now about a dozen of different implementations in the marketplace (e.g., Teradata,[14] Aster Data,[15] Netezza,[16] Vertica,[17] ParAccel,[18] Greenplum[19]). The main aim of these systems is to improve performance through the parallelization of various operations, such as loading data, building indices and

---

[14]http://www.teradata.com/.

[15]http://www.asterdata.com/.

[16]http://www.netezza.com/.

[17]http://www.vertica.com/.

[18]http://www.paraccel.com/.

[19]http://www.greenplum.com/.

evaluating queries. These systems are usually designed to run on top of a shared-nothing architecture [38], where data may be stored in a distributed fashion and input/output speeds are improved by using multiple CPUs and disks in parallel. On the other hand, there are some key reasons that make MapReduce a more preferable approach over a parallel RDBMS in some scenarios [10], such as:

- Formatting and loading a huge amount of data into a parallel RDBMS in a timely manner is a challenging and time-consuming task.
- The input data records may not always follow the same schema. Developers often want the flexibility to add and drop attributes, and the interpretation of an input data record may also change over time.
- Large scale data processing can be very time consuming and therefore it is important to keep the analysis job going even in the event of failures. While most parallel RDBMSs have fault tolerance support, a query usually has to be restarted from scratch, even if just one node in the cluster fails. In contrast, MapReduce deals more gracefully with failures and can redo only the part of the computation that was lost because of a failure.

There has been a long debate on the comparison between the MapReduce framework and parallel database systems[20] [39]. Pavlo et al. [34] have conducted a large scale comparison between the Hadoop implementation of the MapReduce framework and parallel SQL database management systems, in terms of performance and development complexity. The results of this comparison have shown that parallel database systems displayed a significant performance advantage over MapReduce in executing a variety of data intensive analysis tasks. On the other hand, the Hadoop implementation was significantly easier and more straightforward to set up and use in comparison to that of the parallel database systems. MapReduce have also shown to have superior performance in minimizing the amount of work that is lost when a hardware failure occurs. In addition, MapReduce (with its open source implementations) represents a very cheap solution in comparison to the expensive parallel DBMS solutions [39].

The *HadoopDB* project[21] is a hybrid system that tries to combine the scalability advantages of MapReduce with the performance and efficiency advantages of parallel databases [1]. The basic idea behind HadoopDB is to connect multiple single node database systems (PostgreSQL) using Hadoop as the task coordinator and network communication layer. Queries are expressed in SQL but their execution is parallelized across nodes using the MapReduce framework; however, as much of the single node query work as possible is pushed inside of the corresponding node databases. Thus, HadoopDB tries to achieve fault tolerance and the ability to operate in heterogeneous environments by inheriting the scheduling and job tracking implementation from Hadoop. Parallely, it tries to achieve the performance of parallel databases by doing most of the query processing inside the database engine.

---

[20]http://databasecolumn.vertica.com/database-innovation/mapreduce-a-major-step-backwards/.

[21]http://db.cs.yale.edu/hadoopdb/hadoopdb.html.

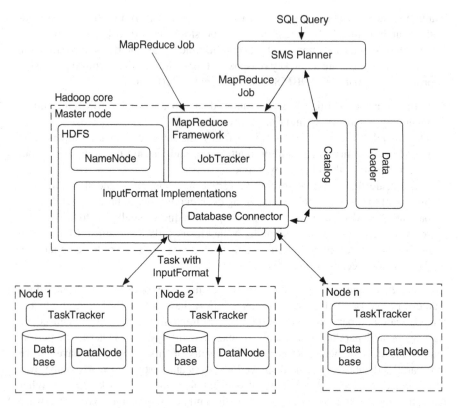

Fig. 1.20 The architecture of HadoopDB [1]

Figure 1.20 illustrates the architecture of HadoopDB, which consists of two layers: (1) a data storage layer or the Hadoop Distributed File System[22] (HDFS), and (2) a data processing layer or the MapReduce Framework. In this architecture, HDFS is a block-structured file system managed by a central *NameNode*. Individual files are broken into blocks of a fixed size and distributed across multiple *DataNodes* in the cluster. The NameNode maintains metadata about the size and location of blocks and their replicas. The MapReduce Framework follows a simple master-slave architecture. The master is a single *JobTracker* and the slaves or worker nodes are *TaskTrackers*. The JobTracker handles the runtime scheduling of MapReduce jobs and maintains information on each TaskTracker's load and available resources. The *Database Connector* is the interface between independent database systems residing on nodes in the cluster and TaskTrackers. The Connector connects to the database, executes the SQL query and returns results as key-value pairs. The *Catalog* component maintains metadata about the databases, their location, replica locations

---

[22]http://hadoop.apache.org/hdfs/.

and data partitioning properties. The *Data Loader* component is responsible for globally repartitioning data on a given partition key upon loading and breaking apart single node data into multiple smaller partitions or chunks. The *SMS planner* extends the HiveQL translator [40] and transforms SQL into MapReduce jobs that connect to tables stored as files in HDFS. Abouzeid et al. [2] have demonstrated HadoopDB in action running two different application types: (1) a semantic web application that provides biological data analysis of protein sequences, and (2) a classical business data warehouse.

Teradata [44] has recently started to follow the same approach of integrating Hadoop and parallel databases. It provides a fully parallel load utility to load Hadoop data to its datawarehouse store. Moreover, it provides a database connector for Hadoop, which allows MapReduce programs to directly access Teradata datawarehouses' data via JDBC drivers without the need of any external steps of exporting (from DBMS) and loading data to Hadoop. It also provides a *Table* user-defined function which can be called from any standard SQL query to retrieve Hadoop data directly from Hadoop nodes in parallel. This means that any relational tables can be joined with the Hadoop data that are retrieved by the Table UDF, and any complex business intelligence capability provided by Teradata's SQL engine can be applied to both Hadoop data and relational data. Hence, no extra steps of exporting/importing Hadoop data to/from the Teradata datawarehouse are required.

1.6   Case Studies

MapReduce-based systems are increasingly being used for large-scale data analysis. There are several reasons for this [28], such as:

- *The interface of MapReduce is simple yet expressive.* Although MapReduce only involves two functions, map and reduce, a number of data analytical tasks, including traditional SQL query, data mining, machine learning and graph processing, can be expressed with a set of MapReduce jobs.
- *MapReduce is flexible.* MapReduce is designed to be independent of storage systems and is able to analyze various kinds of data, structured and unstructured.
- *MapReduce is scalable.* An installation of MapReduce can run over thousands of nodes on a shared-nothing cluster, while keeping to provide fine-grain fault tolerance whereby only tasks on failed nodes need to be restarted.

The above-mentioned advantages have triggered several research efforts that aim at applying the MapReduce framework for solving challenging data processing problems on large scale datasets in a wide spectrum of domains. For example, *Mahout*[23] is an apache project which is designed with the aim of building scalable machine learning libraries using the MapReduce framework. *Ricardo* [15] is a

---

[23]http://mahout.apache.org/.

scalable platform for applying sophisticated statistical methods over huge data repositories. It is designed to facilitate the *trading* between *R* (a famous statistical software) and Hadoop, where each trading partner performs the tasks that it does best. In particular, this trading is performed in a way that *R* sends aggregation-processing queries to Hadoop, while Hadoop sends aggregated data to *R* for advanced statistical processing or visualization.

MapDupReducer [43] is a MapReduce-based system which has been developed for supporting the problem of near duplicate detection over massive datasets. Vernica et al. [42] have proposed an approach to efficiently perform set-similarity joins in parallel using the MapReduce framework. In particular, they have proposed a 3-stage approach for end-to-end set-similarity joins. The approach takes as input a set of records and outputs a set of joined records based on a set-similarity condition. It partitions the data across nodes in order to balance the workload and minimize the need for replication. Morales et al. [21] have presented two matching algorithms, *GreedyMR* and *StackMR*, which are geared for the MapReduce paradigm and aim to distribute content from information suppliers to information consumers on social media applications. In particular, they seek to maximize the overall relevance of the matched content from suppliers to consumers, while regulating the overall activity.

Surfer [14] is a large scale graph processing engine which is designed to execute in the cloud. Surfer provides two basic primitives for programmers: *MapReduce* and *propagation*. In this engine, MapReduce processes different key-value pairs in parallel, and propagation is an iterative computational pattern that transfers information along the edges from a vertex to its neighbors in the graph. In principle, these two primitives are complementary in graph processing where MapReduce is suitable for processing flat data structures (e.g., vertex-oriented tasks), while propagation is optimized for edge-oriented tasks on partitioned graphs.

Lattanzi et al. [30] have presented an approach for solving graph problems using the MapReduce framework. In particular, they present parallelized algorithms for minimum spanning trees, maximal matchings, approximate weighted matchings, approximate vertex and edge covers and minimum cuts. Cary et al. [12] presented an approach for applying the MapReduce model in the domain of spatial data management. In particular, they focus on the bulk-construction of R-Trees and aerial image quality computation, which involves vector and raster data.

Abouzeid et al. [2] have demonstrated that *HadoopDB* in conjunction with a column-oriented database can provide a promising solution for supporting efficient and scalable semantic web applications. Ravindra et al. [36] have presented an approach for parallelizing the processing of analytical queries on RDF graph models. In particular, they extended the function library of *Pig Latin* to include functions that aid in operator-coalescing and look-ahead processing to reduce the I/O costs that arise from repeated processing and materialization of intermediate results.

1.7   Discussion and Conclusions

MapReduce has emerged as a popular way to harness the power of large clusters of computers. Currently, MapReduce serves as a platform for a considerable amount of massive data analysis. It allows programmers to think in a *data-centric* fashion where they can focus on applying transformations to sets of data records, while the details of distributed execution and fault tolerance are transparently managed by the MapReduce framework. Gu and Grossman [25] have reported the following important lessons, which they have learned from their experiments with the MapReduce framework:

- *The importance of data locality.* Locality is a key factor, especially when relying on inexpensive commodity hardware.
- *Load balancing and the importance of identifying hot spots.* With poor load balancing, the entire system can be waiting for a single node. Thus, it is important to eliminate any "hot spots", which can be caused by data access (accessing data from a single node) or network I/O (transferring data into or out of a single node).
- *Fault tolerance comes with a price.* In some cases, fault tolerance introduces extra overhead in order to replicate the intermediate results. For example, in the cases of running on small to medium sized clusters, it might be reasonable to favor performance and re-run any failed intermediate task when necessary.
- *Streams are important.* Streaming is important in order to reduce the total running time of MapReduce jobs.

Jiang et al. [28] have conducted an in-depth performance study of MapReduce using its open source implementation, Hadoop. As an outcome of this study, they identified some factors that can have significant performance effect on the MapReduce framework. These factors are described as follows:

- Although MapReduce is independent of the underline storage system, it still requires the storage system to provide efficient I/O modes for scanning data. The experiments of the study on HDFS show that direct I/O outperforms streaming I/O by 10–15 %.
- MapReduce can utilize three kinds of indices, namely range-indices, block-level indices and database indexed tables, in a straightforward way. The experiments of the study show that the range-index improves the performance of MapReduce by a factor of 2 in the selection task and a factor of 10 in the join task when selectivity is high.
- There are two kinds of decoders for parsing the input records: mutable decoders and immutable decoders. The study claims that only immutable decoders intro- duce performance bottleneck. To handle database-like workloads, MapReduce users should strictly use mutable decoders. A mutable decoder is faster than an immutable decoder by a factor of 10, and improves the performance of selection by a factor of 2. Using a mutable decoder, even parsing the text record is efficient.
- Map-side sorting exerts negative performance effect on large aggregation tasks, which require nontrivial key comparisons and produce millions of groups.

Therefore, fingerprinting-based sort can be used to significantly improve the performance of MapReduce on such aggregation tasks. The experiments show that fingerprinting-based sort outperforms direct sort by a factor of 4–5, and improves overall performance of the job by 20–25%.

- The scheduling strategy affects the performance of MapReduce, as it can be sensitive to the processing speed of slave nodes, and slows down the execution time of the entire job by 25–35% [48].

The experiments of the study show that with proper engineering for these factors, the performance of MapReduce can be improved by a factor of 2.5–3.5, and approaches the performance of Parallel Databases.

In general, to run a single program in a MapReduce framework, a number of tuning parameters (e.g. memory allocation, concurrency, I/O optimization, network bandwidth usage) have to be set by users or system administrators. In practice, users may often run into performance problems because they do not know how to set these parameters. In addition, as MapReduce is a relatively new technology, it is not easy to find qualified administrators. Babu [6] has proposed some techniques to *automate* the setting of tuning parameters for MapReduce programs. The aim of these techniques is to provide good out-of-the-box performance for ad hoc MapReduce programs that run on large datasets. Babu suggested the following research agenda to automatically configure the parameters for MapReduce jobs:

- There is a need to conduct a comprehensive empirical study with a representative class of MapReduce programs and different cluster configurations to understand (and potentially model) parameter impacts, interactions, and response surfaces.
- Developing cost models that are useful to recommend good parameter settings for MapReduce job configuration parameters.
- Tune the performance of a MapReduce program that is run repeatedly (e.g., for daily report generation) and whose current performance is unsatisfactory.
- Developing mechanisms that can automatically generate an execution plan, which is composed of one or more MapReduce jobs for a higher-level operation, like join.

The cluster-level energy management of the MapReduce framework is another interesting research direction. Lang and Patel [29] have investigated the approach to power down (and power up) MR nodes in order to save energy in periods of low utilization. In particular, they compared between two strategies for MR energy management: (1) Covering Set (CS) strategy that keeps only a small fraction of the nodes powered up during periods of low utilization, and (2) All-In Strategy (AIS) that uses all the nodes in the cluster to run a workload and then powers down the entire cluster. The comparison shows that there are two crucial factors that affect the effectiveness of these two methods: (1) the computational complexity of the workload, and (2) the time taken to transition nodes to and from a low power (deep hibernation) state to a high performance state. The comparison evaluation also shows that *CS* is more effective than *AIS* only when the computational complexity of the workload is low (e.g., linear), and that the time it takes for the hardware to transition a node to and from a low power state is a relatively large fraction of the overall workload time (i.e., the workload execution time is small). In all other cases, the *AIS* shows better performance over *CS* in terms of energy savings and response time performance.

We believe that this survey of the MapReduce family of approaches would be useful for the future development of MapReduce-based data processing systems. In addition, we are convinced that there is still room for further optimization and advancement in several directions on the spectrum of the MapReduce framework that is still required to bring forward the vision of providing large scale data analysis as a commodity for novice end-users.

References

1. Abouzeid, A., Bajda-Pawlikowski, K., Abadi, D., Rasin, D.A., Silberschatz, A.: HadoopDB: an architectural hybrid of MapReduce and DBMS technologies for analytical workloads. PVLDB **2**(1), 922–933 (2009)
2. Abouzeid, A., Bajda-Pawlikowski, K., Huang, J., Abadi, D., Silberschatz, A.: HadoopDB in action: building real world applications. In: SIGMOD, Indianapolis, 2010, pp. 1111–1114
3. Afrati, F., Ullman, J.: Optimizing joins in a map-reduce environment. In: EDBT, Lausanne, 2010, pp. 99–110
4. Alvaro, P., Hellerstein, J., Elmeleegy, K., Condie, T., Conway, N., Sears, R.: MapReduce online. In: NSDI, San Jose, 2010
5. Armbrust, M., Fox, A., Rean, G., Joseph, A., Katz, R., Konwinski, A., Gunho, L., David, P., Rabkin, A., Stoica, I., Zaharia, M.: Above the clouds: a Berkeley view of cloud computing, Dept. Electrical Eng. and Comput. Sciences, University of California, Berkeley, Tech. Rep. UCB/EECS, vol. 28, 2009
6. Babu, S.: Towards automatic optimization of MapReduce programs. In: SoCC, Indianapolis, 2010, pp. 137–142
7. Bajda-Pawlikowski, K., Abadi, D., Silberschatz, A., Paulson, E.: HadoopDB in action: efficient processing of data warehousing queries in a split execution environment. In: SIGMOD, Athens, 2011, pp. 1165–1176
8. Bell, G., Gray, J., Szalay, A.: Petascale computational systems. IEEE Comput. **39**(1), 110–112 (2006)
9. Beyer, K., Ercegovac, V., Gemulla, R., Balmin, A., Eltabakh, M., Kanne, C., Ozcan, F., Shekita, E.: Jaql: a scripting language for large scale semistructured data analysis. PVLDB **4**(11), 1272–1283 (2011)
10. Blanas, S., Patel, J., Ercegovac, V., Rao, J., Shekita, E., Tian, Y.: A comparison of join algorithms for log processing in MapReduce. In: SIGMOD, Indianapolis, 2010, pp. 975–986
11. Bu, Y., Howe, B., Balazinska, M., Ernst, M.: HaLoop: efficient iterative data processing on large clusters. PVLDB **3**(1), 285–296 (2010)
12. Cary, A., Sun, Z., Hristidis, V., Rishe, N.: Experiences on processing spatial data with MapReduce. In: SSDBM, New Orleans, 2009, pp. 302–319
13. Chaiken, R., Jenkins, B., Larson, P., Ramsey, B., Shakib, D., Weaver, S., Zhou, J.: SCOPE: easy and efficient parallel processing of massive data sets. PVLDB **1**(2), 1265–1276 (2008)
14. Chen, R., Weng, X., He, B., Yang, M.: Large graph processing in the cloud. In: SIGMOD, Indianapolis, 2010, pp. 1123–1126
15. Das, S., Sismanis, Y., Beyer, K., Gemulla, R., Haas, P., McPherson, J.: Ricardo: integrating R and Hadoop. In: SIGMOD, Indianapolis, 2010, pp. 987–998

16. Dean, J., Ghemawat, S.: MapReduce: simplified data processing on large clusters. In: OSDI, San Francisco, 2004, pp. 137–150
17. Dean, J., Ghemawat, S.: MapReduce: simplified data processing on large clusters. Commun. ACM **51**(1), 107–113 (2008)
18. Dean, J., Ghemawat, S.: MapReduce: a flexible data processing tool. Commun. ACM **53**(1), 72–77 (2010)
19. Dittrich, J., Quiane-Ruiz, J., Jindal, A., Kargin, Y., Setty, V., Schad, J.: Hadoop++: making a yellow elephant run like a cheetah (without it even noticing). PVLDB **3**(1), 518–529 (2010)
20. Eltabakh, M., Tian, Y., Ozcan, F., Gemulla, R., Krettek, A., McPherson, J.: CoHadoop: flexible data placement and its exploitation in Hadoop. PVLDB **4**(9), 575–585 (2011)
21. Francisci Morales, G., Gionis, A., Sozio, M.: Social content matching in MapReduce. PVLDB **4**(7), 460–469 (2011)
22. Friedman, E., Pawlowski, P., Cieslewicz, J.: SQL/MapReduce: a practical approach to self-describing, polymorphic, and parallelizable user-defined functions. PVLDB **2**(2), 1402–1413 (2009)
23. Gates, A., Natkovich, O., Chopra, S., Kamath, P., Narayanam, S., Olston, C., Reed, B., Srinivasan, S., Srivastava, U.: Building a highlevel data ow system on top of MapReduce: the pig experience. PVLDB **2**(2), 1414–1425 (2009)
24. Ghemawat, S., Gobioff, H., Leung, S.: The Google file system. In: SOSP, Bolton Landing, 2003, pp. 29–43
25. Gu, Y., Grossman, R.: Lessons learned from a year's worth of benchmarks of large data clouds. In: SC-MTAGS, Portland, 2009
26. Hey, T., Tansly, S., Tolle, K. (eds.): The Fourth Paradigm: Data-Intensive Scientific Discovery. Microsoft Research, Redmond (2009)
27. Isard, M., Budiu, M., Yu, Y., Birrell, A., Fetterly, D.: Dryad: distributed data-parallel programs from sequential building blocks. In: EuroSys, Lisbon, 2007, pp. 59–72
28. Jiang, D., Chin Ooi, B., Shi, L., Wu, S.: The performance of MapReduce: an in-depth study. PVLDB **3**(1), 472–483 (2010)
29. Lang, W., Patel, J.: Energy management for MapReduce clusters. PVLDB **3**(1), 129–139 (2010)
30. Lattanzi, S., Moseley, B., Suri, S., Vassilvitskii, S.: Filtering: a method for solving graph problems in MapReduce. In: SPAA, San Jose, 2011, pp. 85–94
31. Murray, D., Hand, S.: Scripting the cloud with Skywriting. In: HotCloud, USENIX Workshop, Boston, 2010
32. Nykiel, T., Potamias, M., Mishra, C., Kollios, G., Koudas, N.: MRShare: sharing across multiple queries in MapReduce. PVLDB **3**(1), 494–505 (2010)
33. Olston, C., Reed, B., Srivastava, U., Kumar, R., Tomkins, A.: Pig latin: a not-so-foreign language for data processing. In: SIGMOD, Vancouver, 2008, pp. 1099–1110
34. Pavlo, A., Paulson, E., Rasin, A., Abadi, D., DeWitt, D., Madden, S., Stonebraker, M.: A comparison of approaches to large-scale data analysis. In: SIGMOD, Providence, 2009, pp. 165–178
35. Pike, R., Dorward, S., Griesemer, R., Quinlan, S.: Interpreting the data: parallel analysis with Sawzall. Sci. Program. **13**(4), 277–298 (2005)
36. Ravindra, P., Deshpande, V., Anyanwu, K.: Towards scalable RDF graph analytics on MapReduce. In: MDAC, Raleigh, 2010
37. Sakr, S., Liu, A., Batista, D., Alomari, M.: Hive – a survey of large scale data management approaches in cloud environments. IEEE Commun. Surv. Tutor. **13**(3), 311–336 (2011)
38. Stonebraker, M.: The case for shared nothing. IEEE Database Eng. Bull. **9**(1), 4–9 (1986)
39. Stonebraker, M., Abadi, D., DeWitt, D., Madden, S., Paulson, E., Pavlo, A., Rasin, A.: MapReduce and parallel DBMSs: friends or foes? Commun. ACM **53**(1), 64–71 (2010)
40. Thusoo, A., Sarma, J., Jain, N., Shao, Z., Chakka, P., Anthony, S., Liu, H., Wyckoff, P., Murthy, R.: Hive – a warehousing solution over a map-reduce framework. PVLDB **2**(2), 1626–1629 (2009)

41. Thusoo, A., Sarma, J., Jain, N., Shao, Z., Chakka, P., Anthony, S., Liu, H., Wyckoff, P., Murthy, R.: Hive – a petabyte scale data warehouse using Hadoop. In: ICDE, Long Beach, 2010, pp. 996–1005
42. Vernica, R., Carey, M., Li, C.: Efficient parallel set-similarity joins using MapReduce. In: SIGMOD, Indianapolis, 2010, pp. 495–506
43. Wang, C., Wang, J., Lin, X., Wang, W., Wang, H., Li, H., Tian, W., Xu, J., Li, R.: MapDupReducer: detecting near duplicates over massive datasets. In: SIGMOD, Indianapolis, 2010, pp. 1119–1122
44. Xu, Y., Kostamaa, P., Gao, L.: Integrating Hadoop and parallel DBMS. In: SIGMOD, Indianapolis, 2010, pp. 969–974
45. Yang, H., Parker, D.: Traverse: simplified indexing on large map-reduce-merge clusters. In: DASFAA, Brisbane, 2009, pp. 308–322
46. Yang, H., Dasdan, A., Hsiao, R., Parker, D.: Map-reduce-merge: simplified relational data processing on large clusters. In: SIGMOD, Beijing, 2007, pp. 1029–1040
47. Yu, Y., Isard, M., Fetterly, D., Budiu, M., Erlingsson, U., Gunda, P., Currey, J.: DryadLINQ: a system for general-purpose distributed data-parallel computing using a high-level language. In: OSDI, San Diego, 2008, pp. 1–14
48. Zaharia, M., Konwinski, A., Joseph, A., Katz, R., Stoica, I.: Improving MapReduce performance in heterogeneous environments. In: OSDI, San Diego, 2008, pp. 29–42
49. Zhou, J., Larson, P., Chaiken, R.: Incorporating partitioning and parallel plans into the SCOPE optimizer. In: ICDE, Long Beach, 2010, pp. 1060–1071

Chapter 2
Optimization of Massively Parallel Data Flows

Fabian Hueske and Volker Markl

Abstract Massively parallel data analysis is an emerging research topic that is motivated by the continuous growth of data sets and the rising complexity of data analysis tasks. To facilitate the analysis of big data, several parallel data processing frameworks, such as MapReduce and parallel data flow processors, have emerged. However, the implementation and tuning of parallel data analysis tasks requires expert knowledge and is very time-consuming and costly. Higher-level abstraction frameworks have been designed to ease the definition of analysis tasks. Optimizers can automatically generate efficient parallel execution plans from higher-level task definitions. Therefore, optimization is a crucial technology for massively parallel data analysis. This chapter presents the state of the art in optimization of parallel data flows. It covers higher-level languages for MapReduce, approaches to optimize plain MapReduce jobs, and optimization for parallel data flow systems. The optimization capabilities of those approaches are discussed and compared with each other. The chapter concludes with directions for future research on parallel data flow optimization.

2.1 Introduction

Today, many companies and research groups from various domains are facing challenges due to two ongoing trends. First, the amount of data that needs to be processed grows at unpreceded rates [60]. Second, the depth of data analysis steadily increases, i.e., analysis tasks become more complex [25]. Companies and facilities that are affected by these trends come from the Internet business, biology, climate, or astronomy research, just to name a few [21]. The sheer size of the data sets prohibits the use of centralized systems and requires distributed hardware

F. Hueske (✉) • V. Markl
Technische Universität Berlin, Berlin, Germany
e-mail: fabian.hueske@tu-berlin.de; volker.markl@tu-berlin.de

A. Gkoulalas-Divanis and A. Labbi (eds.), *Large-Scale Data Analytics*,
DOI 10.1007/978-1-4614-9242-9_2, © Springer Science+Business Media New York 2014

architectures. Shared-nothing systems based on many commodity servers are very scalable and provide impressive I/O bandwidth due to the large number of hard disks. However, software frameworks running on such architectures must be capable of handling node failures which become very common at large scales.

In 2004, Google promoted a software stack consisting of MapReduce [26] and the Google File System (GFS) [35] to process huge amounts of data on large clusters of commodity hardware. MapReduce offers a simple interface to implement parallel data analysis tasks. Most aspects that make parallel software development challenging, i.e., communication, scheduling, and fault-tolerance, do not need to be handled by the programmer. Instead the MapReduce execution framework takes care of those issues. Since its publication, the MapReduce paradigm has become very popular. The Apache Hadoop project [7] provides open-source implementations of MapReduce and GFS.

MapReduce differs from traditional approaches to manage and analyze large data sets. David DeWitt and Micheal Stonebraker argue that MapReduce is "a major step backwards" because it neglects 40 years of database research. Specifically, they criticize the lack of schema information, sophisticated data access methods and query processing algorithms, a declarative language interface, and a query optimizer. Those features had significant share in the success of relational database systems. However, the research and open-source communities, as well as industry, have started numerous efforts to improve MapReduce in many aspects, e.g., by building on top of MapReduce, extending it, or providing advanced MapReduce algorithms.

While using MapReduce as a paradigm for massively parallel data analytics has been very popular, alternative systems have been built [15, 16, 45]. These frameworks process arbitrary acyclic data flows in a parallel fashion and incorporate many techniques of traditional parallel database systems. In contrast to MapReduce which includes a fixed execution model, the execution layer of parallel data flow processors is much more flexible.

However, MapReduce as well as parallel data flow systems have in common that implementation and tuning of massively parallel data analysis tasks requires expert knowledge and is very time-consuming and costly. Several approaches, such as higher-level languages and parallel programming models, have been proposed to ease the definition of complex parallel data analysis tasks. Tasks defined on such a level of abstraction must be compiled into an execution plan in order to be processed by an execution framework. Optimizers automatically generate plans to efficiently execute complex data analysis tasks at large scales. Therefore, optimization is a crucial aspect for massively parallel data analysis.

This chapter discusses the state-of-the-art in massively parallel data flow optimization. Although parallel *Relational Database Management Systems* (RDBMSs) process parallel data flows, we do not cover this well-researched domain and focus on current approaches, such as MapReduce and Dryad. Many approaches published in recent years target the popular MapReduce programming and execution model. Much work was proposed to improve MapReduce itself or to integrate techniques which are well-known from the domain of relational databases. Some of these approaches increase the optimization potential of MapReduce. Due to the amount of publications in this context, we restrict the scope of the chapter to frameworks that provide a user front end to specify data analysis tasks, an optimizer, and an execution engine to process the optimized task.

Besides work targeting the popular MapReduce model, a couple of alternative systems for large scale data analysis have been built. These systems are based on massively parallel data flow engines. Due to their flexibility to process arbitrary acyclic data flows, these systems are well suited for optimization. On top of these flexible execution engines, innovative approaches to specify data analysis tasks have been developed that incorporate sophisticated optimizers capable of many optimizations known from parallel relational database systems.

The chapter is structured as follows. Section 2.2 recapitulates the basics of query optimization in parallel relational database systems and the MapReduce programming and execution models. The next sections present the state-of-the-art in optimization of massively parallel data flows. Section 2.3 discusses higher-level languages for MapReduce and Sect. 2.4 presents approaches to optimize plain MapReduce jobs. In Sect. 2.5 parallel data flow systems, their programming interfaces, and optimization capabilities are discussed. A summary and comparison of the presented approaches is given in Sect. 2.6. Finally, Sect. 2.7 proposes areas of future work and concludes the chapter.

2.2 Background

This section provides a brief overview of well-known techniques relevant to the topics discussed in this chapter. We shortly recapitulate optimization in parallel relational database systems in Sect. 2.2.1 and give an introduction into the MapReduce paradigm in Sect. 2.2.2.

2.2.1 Query Optimization in Parallel RDBMSs

Query optimization is one of the most extensively researched topics in the field of information management. The popularity of SQL and the relational data model necessitated and powered the advances in database optimization. In relational database systems, the optimization potential rises from the relational algebra and the declarative query language SQL. Rather than imperatively specifying what should be done, SQL queries define the desired result. Declarative queries give a database system the freedom to decide how to compute the requested result. Since there are many ways to execute a single query, the goal of query optimization is to find the most efficient execution plan for a given query. We first give a high-level overview of database query optimization in general and then discuss issues that emerge in the context of parallel database systems.

Most modern database optimizers follow a *three-staged approach* [40]. After the query is parsed, syntactically checked, and compiled into an internal representation, logical optimization is performed. This is done by logically rewriting the internal query representation. Common rewrite rules are selection and projection push-down, incorporation of views and un-nesting of subqueries. Since these rules are expected to generally improve query performance, they are applied when possible.

The third step is physical optimization and generates a physical execution plan. This process includes the choice of physical operators and access paths, and determines the order of joins.[1] There are two common approaches to enumerate plans in physical optimization: *bottom-up* and *top-down*. Bottom-up approaches enumerate candidate execution plans by starting at the base relations and finishing at the root of the execution plan. Commonly, dynamic programming techniques are used to construct an optimal plan based on its optimal subplans [51,58]. In contrast, top-down enumeration techniques start at the root of an execution plan and move towards the base relations of a query. Branch and bound algorithms are often used to prune large fractions of the search space [32, 36]. All techniques rely on the ability to estimate the cost of a subplan. Cost estimates are used to compare equivalent subplans and to prune the search space. Therefore, accurate estimates are crucial for quality and performance of the optimization.

The cost model is a collection of functions that compute a cost metric, such as resource consumption or response time. Input parameters of these functions include statistics of the input data (e.g., cardinalities, distributions), query properties (e.g., local and join predicates), and the amount of available resources (e.g., memory and CPU). The quality of the cost estimates significantly depends on the accuracy of the input data statistics. The optimizer retrieves such information from a component, called *catalog*, which stores metadata, such as table schemas and data statistics.

Query optimizers for shared-nothing parallel database systems must take several additional aspects into account that blow the search space [27]. Parallel execution plans consist of subplans that are concurrently executed on multiple machines and linked by data reorganization steps, such as partitionings. Multiple database instances process a query in parallel and ship partial results over the network in order to compute the final result. Contrary to centralized systems, where costs for disk I/O are predominant, network I/O costs account for the biggest portion of the absolute costs in parallel database systems. Therefore, parallel cost models must take shipping costs and concurrent execution into account [47].

In order to reduce the amount of data shipped over the network, several techniques have been proposed. Data placement strategies, such as partitioning, replication and co-location, enable the local computation of expensive operations, such as joins and aggregations [47]. In many cases, large fractions of a query can be processed without any network communication. Reasoning about existing partitionings can significantly reduce the amount of shipped data and improve the

---

[1]The optimal join order depends on the choice of the physical operators. Therefore, join ordering is done as part of the physical optimization, although it is a logical rewrite.

query runtime. Further techniques to reduce network traffic are semi-join reduction and partial aggregation. Semi-join reduction improves distributed joins and reduces relations prior to shipping, by filtering out tuples which will not match in the join [17, 47]. Partial aggregation is similar to the MapReduce's Combiner and reduces shipped data sets by locally applying partial aggregation functions.

Most database systems follow one of two approaches to incorporate parallelism into their optimizer. Either, the characteristics of the parallel execution environment are integrated into an existing optimizer framework or an external parallelizer is called after the query was optimized for a centralized environment. Due to lack of space, we do not discuss the different approaches in detail here. Reference [52] gives a detailed list of work in this context.

2.2.2 MapReduce

Google proposed MapReduce as an approach to process huge amounts of data in a massively parallel fashion [26]. This paper introduces MapReduce as a programming model and an execution model. Since its publication, MapReduce has gained a lot of attraction in research, industry and the open-source community. A couple of MapReduce frameworks have been implemented, among them the popular Apache Hadoop framework [7]. Numerous MapReduce algorithms, extensions, and components that are based on Hadoop have been published.

This section provides a brief overview of MapReduce. We start with the MapReduce programming model and continue with the execution model. Finally, MapReduce's strengths and weaknesses, with respect to optimization, are discussed.

2.2.2.1 MapReduce Programming Model

The parallel MapReduce programming model is inspired by functional programming and is very simple in its essence [26]. The programming model is based on the principles of data parallelism, which enables processing at large scale. Data is distributed over multiple compute nodes, each working on a subset of the data.

MapReduce's data model are pairs of keys and values which can be of any user-defined type. Atomic types, such as integer values, are supported as well as complex data structures, e.g., hash tables or nested structures. MapReduce programs consist of two second-order functions, called *Map* and *Reduce*, each processing a bag of key/value pairs. A program is specified by implementing two first-order user functions, one for Map and one for Reduce. These implementations can perform arbitrary computations. The second-order functions Map and Reduce call their first-order functions with subsets of their input. All calls of user functions are independent from each other, which is the prerequisite for data parallelism. Map and Reduce differ in the way they build the subsets of key/value pairs to call the user code. Map calls its first-order function for each single key/value pair

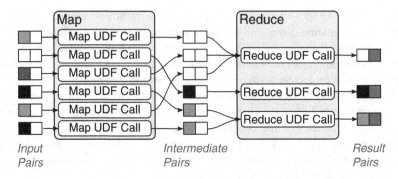

Fig. 2.1 The MapReduce programming model

separately. In theory, each pair could be processed on a dedicated machine due to the independence of user code calls. Reduce calls its user function with a group of key/value pairs, where a group consists of all pairs that share the same key. Both user functions can modify the type and value of keys and values and emit none, one or multiple pairs.

A MapReduce program has a fixed structure that is shown in Fig. 2.1. A data source reads the input data and generates key/value pairs. Each pair is independently processed by Map. Pairs resulting from Map are handed to Reduce which processes them group-wise per key. Finally, the result of the Reduce function is emitted and the program finishes.

2.2.2.2 MapReduce Execution Model

Along with the programming model, Dean et al. [26] introduce a model to execute MapReduce programs at large scale in a fault tolerant fashion. The architectural stack founds on a distributed file system, such as GFS [35]. The files stored in the distributed file system are split into fixed-sized chucks. The chunks are distributed and replicated among the nodes that belong to the file system cluster. Distribution provides high performance parallel scans. Replication, on the other hand, ensures fault tolerance and availability.

Figure 2.2 depicts the execution of a MapReduce job. A MapReduce job is split into m Map and n Reduce tasks. Each task is scheduled to a compute node of the cluster and processes a subset of the overall data. The m Map tasks read chunks of the input file in parallel from the distributed file system. A task generates key/value pairs from its chunk and hands them individually to the Map user function. All pairs that are emitted from the Map function are collected. The pairs are partitioned into n buckets using a deterministic partitioning function that is applied to the key. Each bucket is written to the disk of the Map node. The i-th Reduce task collects from all Map nodes their i-th buckets, pulls them over the network, and sorts them by key in

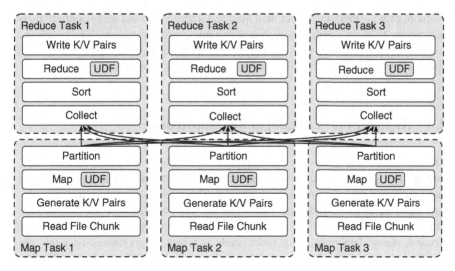

Fig. 2.2 The MapReduce execution model

order to obtain a sequence of key/value pairs which is grouped by key.[2] The Reduce task calls the Reduce user function for each key-group and writes all pairs emitted by the function to the distributed file system. The job is finished after all Map and Reduce tasks are finished.[3]

2.2.2.3 MapReduce and Optimization

The popularity of the MapReduce paradigm is due to its simple programming interface and the highly scalable and fault-tolerant execution model. In contrast to parallel database systems, MapReduce frameworks are much easier to setup and maintain, and enable ad-hoc analysis of large data sets.

These properties, however, pose several challenges for optimization as it is known from relational database systems. MapReduce jobs are specified using user-defined functions which impedes reasoning about the semantics of a job. The lack of schema and data statistics prohibits most cost-based optimization approaches. The usage of a distributed file system for data storage complicates the application of sophisticated storage and data placement approaches, such as indexes and data co-location. Finally, the fixed execution strategy reduces the degrees of freedom for physical optimization. However, some MapReduce systems, such as

---

[2]Hadoop's implementation varies from the original paper by performing partial sorts already within the Map task. Subsequently, the Reduce task merges the sorted buckets.

[3]Due to lack of space, we do not explain the execution of the optional Combiner. Instead, we refer the reader to the original paper [26].

Hadoop, provide more interfaces than Map and Reduce [28] and many configuration parameters [13]. In combination with alternative implementations for operations, such as for joins, these options span the search space for physical optimization. The approaches presented in the following sections address subsets of these challenges.

2.3 Higher-Level Languages for MapReduce

The motivation for higher-level languages on top of MapReduce is multifold. First, writing data processing tasks with MapReduce requires programming experience and familiarity with the programming model. This is even more true for the implementation of efficient programs and tasks that do not nicely fit the MapReduce model. The development of such complex tasks is often very time consuming. Second, many data processing tasks, such as filters, aggregations or joins, are commonly used. Therefore, reuse of existing code is highly advisable. Finally, many users of large-scale analytical platforms are analysts with a limited technical background.

Higher-level languages provide well defined data models and implementations of a set of common operations and allow non-technical persons to analyze massive data sets. Furthermore, they significantly speed up development of analysis tasks due to their high-level abstraction and re-usability.

Since MapReduce's introduction, a couple of higher-level languages have been proposed. Among them are Pig, Hive, Jaql, Sawzall and Cascading. Cascading differs from the others by not providing a language interface. Instead, it offers a Java API to define higher-level data flows, which are compiled to MapReduce without applying any optimization [23]. Sawzall is a low-level language proposed by Google [57]. It provides a scripting interface to ease the definition of MapReduce jobs rather than enabling the specification of complex analysis tasks. This section focuses on Pig, Hive and Jaql, which are semantically rich languages and perform actual optimization. These languages differ in their data models, supported operations, extensibility and additional features. All have in common that their queries are optimized and compiled to MapReduce programs

In the following, we present Pig, Hive, and Jaql. Subsequent to that introduction, we point out the differences between optimizing higher-level languages for MapReduce and SQL for parallel DBMSs. We conclude this section discussing optimizations that are applied to generate efficient MapReduce programs.

2.3.1 Overview of Higher-Level Languages

This section gives a short overview of prominent higher-level languages for the MapReduce programming model. We present their data and operator models and discuss their distinguishing features.

2.3.1.1 Pig

The Apache Pig [10] project develops a platform for analyzing and processing large data sets. Pig was started at Yahoo! Research [34, 55] and became an open-source project in September 2008. Since then, it evolved to one of the most recognized projects in Apache's Hadoop environment. The platform features a scripting language for data analysis and a compiler that translates scripts into Hadoop MapReduce jobs.

Pig operates on a semi-structured and nested data model. Atomic and composed data types, such as numeric values, text, records, bags and associative maps, are supported. Data types can be arbitrarily nested. Many of Pig's operators are known from relational algebra or operate on nested data. The operators have been chosen with a focus on parallel execution. Therefore, non-equi-joins and correlated subqueries are not supported. However, Pig can be easily extended by user-defined operators. Pig follows a procedural programming style, i.e., scripts are composed step by step. Each step performs a single or very few simple transformations or data operations such as selection, grouping, or joining. The procedural style is an obvious contrast to SQL's declarative style. Another difference to SQL is that Pig scripts can compute more than one result set. Due to its procedural programming style, extensive support for user-defined operations and handling of semi-structured data, Pig is well suited for complex data flow definitions, such as ETL processes.

2.3.1.2 Hive

Hive [8, 62, 63] is a data warehouse solution on top of Apache Hadoop. It was designed at Facebook and joined the Apache Hadoop ecosystem as a second higher-level language next to Pig. The first public version was released in April 2009.

Hive stores data as tables in Hadoop's file system HDFS. Tables are organized in partitions and smaller units, called *buckets*. Similar to Pig, Hive's data model consists of atomic types, maps, lists and struct types. Arbitrary nesting of types is supported, as well as user-defined data types. In contrast to Pig, Hive features a system catalog, called *Metastore*, which holds schema and partitioning information of the data imported and managed by Hive. Hive's query language resembles SQL and follows its declarative programming style, which helps users who are familiar with SQL. A common feature of Hive and Pig is the ability to chain and branch queries. This can be used to compute more than one result. The Metastore and the SQL-like language are Hive's main features that qualify it as a large-scale data warehouse.

2.3.1.3   Jaql

Jaql [18] is a declarative scripting language to analyze semi-structured data. It was designed by IBM Research, became an open-source project in 2009, and is integrated into two IBM Big Data analysis products.

Jaql's data model is based on the JavaScript Object Notation (JSON) data model. JSON is semi-structured and nested. Jaql handles data with partial schema information, i.e., Jaql processes schema-free data but in the presence of schema information certain optimizations and more efficient processing techniques can be applied. The language was inspired by XQuery, Pig, Hive and DryadLINQ. Jaql supports operations such as transform, filter, expand, join and merge. In addition, block and if-then-else control structures and error controlling operators, such as catch, fence and timeout, are included. Jaql's syntax follows a step-by-step style similar to Pig, but resembles UNIX pipes for improved readability.

Jaql is built on concepts from functional programming, such as lazy evaluation and higher-order functions. These techniques are responsible for its excellent extensibility and re-usability. A feature that differentiates Jaql from Pig and Hive is physical transparency. Low-level physical operators and high-level declarative operators are conceptually the same. Both types of operators can be mixed when writing Jaql scripts. This allows users to force the execution of certain strategies. In addition, the result of optimization is a valid Jaql script, which significantly eases debugging.

2.3.2   Comparing Higher-Level Languages and RDBMSs

Higher-level languages, such as Pig and Hive, provide an abstraction from the MapReduce programming model. Data processing tasks are specified as queries or scripts. In order to execute those tasks on a MapReduce framework, they are compiled into MapReduce programs. Since most MapReduce frameworks use a fixed execution strategy, the compile step is very similar to query optimization in relational database systems. As discussed in Sect. 2.2.1, query optimizers aim to choose the best performing execution plan among multiple candidates. The optimization of higher-level languages for MapReduce differs in several aspects from query optimization in relational database systems.

First, the presented languages have different characteristics than SQL. This is mainly due to their semi-structured and nested data models and their focus on extensibility. Some languages support the seamless integration of MapReduce jobs which gives more flexibility than user-defined functions (UDFs) in traditional database systems. Since the semantics of user-written data operations are hard to reason about, queries of higher-level languages that use these features have only limited optimization potential, i.e., holistic optimization that spans UDFs can not be performed.

Relational database systems use a storage layer that enforces a strict schema. In contrast, MapReduce approaches store data as files in a distributed file system. The flexibility of file system storage and the support of schema-free data are often listed among the key advantages of MapReduce over traditional database systems. However, schema information is very important for many optimizations, especially for storage organization and data access techniques. Furthermore, lack of schema limits the amount of statistical information on data sets which is required for most cost-based optimizations.

Finally, execution plans generated for relational database systems are complex trees of physical data processing operators. In contrast, generated MapReduce programs must obey the very strict structure of the programming model. Even rather simple queries that perform a join and an aggregation are compiled into multiple depending MapReduce jobs.

2.3.3   Optimization of Higher-Level Languages

This section discusses the optimization of higher-level languages for MapReduce and the compilation into sequences of MapReduce jobs. The set of supported optimizations varies among the presented languages. This is due to the characteristics of the languages, the availability of additional information such as meta data, and the maturity of the project. All presented languages have been released as open-source and are constantly improved and extended.

2.3.3.1   Logical Optimization and Rewrite Heuristics

Pig, Hive, and Jaql include declarative operators with clearly defined semantics. Therefore, logical optimization can be applied on tasks specified in any of these languages. Although, they differ in syntax and programming style, all feature a subset of the extended relational algebra, such as selection, join, grouping and aggregation operators. Logical rewrite rules and heuristics known from relational database optimizers are applicable in their context. The fact that Pig follows a procedural programming style and Jaql a notion similar to UNIX pipes does not impede such logical optimization.

At the current state, Pig [34], Jaql [18], and Hive [63] feature transformation-based optimizers that apply heuristic rewrite rules. Depending on the concrete system, those rules include selection and projection push-down, variable and function inlining, nesting, un-nesting and field access rewrites. For example, the Jaql optimizer features more than 100 rules which are greedily applied. All three projects plan to implement a cost-based optimizer in the future.

Extensibility is among the key features of Jaql, Pig, and Hive. Logical optimizations cannot be applied if the semantics of an affected operator are unknown. Therefore, user-defined operations can limit the applicability of logical transformations. Even if the semantics of a user-defined operation are known, certain rewrite rules can have a negative effect. For example, pushing down an expensive selection predicate can result in high execution costs, since the predicate might be evaluated very often. Therefore, usage of extensibility features can impede certain optimizations. In order to preserve the optimization capabilities for extensions, Jaql provides explicit annotations which reveal characteristics of user code that can be exploited by the optimizer.

2.3.3.2 Physical Optimization

Physical optimization is performed after logical optimization and generates a physical execution plan from a logical query plan. The challenge is to find the best performing execution plan among all valid plans. The performance difference of execution plans is due to alternative local execution and shipping strategies.

Most traditional database systems follow a cost-based approach for physical optimization. Cost-based optimization has three main requirements. First, a cost model that computes for an execution plan and relevant additional input parameters a metric. The optimizer uses that metric to compare competing execution plans. Second, relevant inputs for the cost model, such as data statistics (e.g., input sizes, cardinalities and value distributions) and available resources are necessary to compute the comparing metric. Finally, alternative execution strategies are an essential part of the optimizers search space.

It is important to note that none of the presented languages features a cost-based optimizer [18, 54, 63]. This is due to several reasons. A fully fledged cost model for any of the presented languages would need to represent the performance characteristics of the MapReduce execution framework. For Hadoop such a cost model was developed [41] but has not been included in one of the languages.

Cost functions compute costs based on available system resources and certain estimations, such as the size of intermediate results and the number of distinct values. A starting point for these estimates is statistical information on the input data. This information is propagated through the query plan, i.e., output estimates of an operator are based on output estimates of its preceding operators. The optimizer computes the estimates of intermediate results based on the semantics of the emitting operation, such as filter or join. Current MapReduce-based systems do not collect any statistical information. Hive's catalog Metastore stores schema definitions and partitioning information but does not hold any statistics in the current version [63]. In addition, it is very hard to derive the semantics of user-defined operations, which renders the computation of estimates often impossible. Due to the propagation effect, UDFs usually brake the computation chain of estimates. Therefore, cost-based optimization without input statistics and operator semantics is a very tough problem.

Another issue is the static structure of the MapReduce programming model, which complicates the implementation of certain alternative execution strategies. Nonetheless, a couple of alternative strategies have been developed. Among them are join strategies [19], indexes [28] and columnar storage layouts [48]. However, only a subset of such improvements has been integrated into the presented languages. None of them features indexes or columnar storage layouts.

Instead of cost-based optimization, heuristics are used to compile logically rewritten queries into MapReduce jobs. A commonly applied rule is to minimize the number of MapReduce jobs. Usually, this heuristic produces good results since each MapReduce job comes with high overhead costs resulting from the fixed execution pipeline of reading the input, sorting, shuffling and writing the result. A technique to reduce the number of MapReduce tasks is chaining of record-wise computations within a single Map task [54]. Hive transforms multiple binary joins on the same attribute into a multiway join [62] and automatically chooses the input sides of Reduce-side joins to reduce the amount of data to be held in memory [63]. A more sophisticated approach to process multiple joins within a single Reducer was presented by Ullman et al. [2].

Traditional join-order optimization is currently not applied by higher-level language optimizers due to lack of precise statistical information. Gates et al. [34] state that only few Pig scripts contain more than one join because of Pig's nested data models.

Alternative execution strategies, such as Map-side and Reduce-side joins, can significantly reduce runtimes if chosen in the right context. To overcome the lack of a cost-based optimizer, Pig, Jaql and Hive provide special user hints and leave it to the programmer to explicitly choose the strategy to be used. Pig lets the user pick from three join algorithms, namely a hash-based join, a sort-merge-based join and a fragment-and-replicate join [34]. Hive offers even more hints to control the execution. The programmer can specify to use hash-based partial aggregations and map-side joins, and instruct special handling of skewed data sets [63]. In some contexts the programmer also has to specify execution parameters, such as the amount of main memory for hash-tables.

In contrast to Pig and Hive, Jaql's optimizer performs source-to-source compilation, i.e., the result of the optimization process are valid Jaql scripts that consist of lower-level operators. Such a lower-level script is equivalent to a physical execution plan. This feature is called *physical transparency* and has major implications. First, users can directly use lower-level operators in order to fix the execution strategy, which offers even more control than hints. Second, new lower-level operators can be easily integrated and used without adapting the optimizer. Finally, source-to-source compilation significantly eases debugging.

2.3.3.3 Storage Optimization

Parallel shared-nothing database systems typically use horizontal partitioning to distribute data over all nodes of the system. The choice of the partitioning attribute and function (e.g., hash, range or round-robin partitioning) enables various optimizations and significantly affects the performance of queries. Among the possible optimizations are reduction of scanned partitions due to filter predicates, directed

a

Relation	Partition(Date)	FS Path	DFS Blocks
Sales	2011-01	/rel_Sales/part_2011-01	A, B
	2011-02	/part_2011-02	C
	2011-03	/part_2011-03	D
	2011-04	/part_2011-04	E, F
	2011-05	/part_2011-05	G
	2011-06	/part_2011-06	H, I, J

b

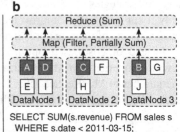

SELECT SUM(s.revenue) FROM sales s
WHERE s.date < 2011-03-15;

Fig. 2.3 Hive partitioning feature: (**a**) Hive metastore and (**b**) Hive query execution

and co-located joins, and local aggregations. By default, MapReduce systems store data as files in distributed file systems. Large files are split into blocks, which are replicated and distributed among nodes. This behavior basically resembles a round-robin or random partitioning and does not provide any semantics that could be used by an optimizer. Due to the transparent data placement of the distributed file system, data co-locations can not be established.

Hive features horizontally partitioned storage, which is implemented on top of Hadoop's distributed file system. Hive splits tables into partitions and buckets, and organizes these in a dictionary hierarchy within the file system. Partitioning information is stored along with schema information in Hive's catalog. During query optimization, Hive checks the catalog and limits table scans to the required set of partitions and buckets [62, 63]. Figure 2.3 depicts Hive's partitioning feature.

Indexes and columnar storage schemes are well-known techniques to improve data access in relational database systems. A couple of approaches to integrate these techniques into MapReduce-based systems have been published [28, 33, 48]. References [31, 54] discuss advanced data placement optimizations that are applicable in the context of MapReduce higher-level languages. Among them are automatic replication of heavily used files, data co-location, and load balancing by separating files that are seldom processed together. Another proposed technique is to store and reuse the results of common operations. This strategy is especially beneficial if some operations are repeatedly computed over unmodified input data. The presented approaches have been extensively researched in the context of relational database systems [3, 50]. However, none of these approaches has been integrated into any of the presented languages.

2.3.3.4 Multi-result Optimization

Pig and Hive support the specification of multiple results within a single script or query. This is different from SQL, where a query defines a single result. Since full scans of the data are very expensive and intermediate results are always materialized by the MapReduce execution framework, both systems consider results that can be reused. Pig's optimizer tries to minimize the computation effort for queries with multiple results by multiplexing computations within a MapReduce job [34]. However, this approach can lead to performance degradation if the processed data exceeds the available main memory. Pig's optimizer is not able to reason about this aspect. Instead, multiplexing is applied by default and the programmer need to manually split multi-output scripts in order to avoid it. Hive applies similar optimizations to reduce the number of MapReduce jobs for multi-output queries [63].

File scans and intermediate results can also be shared by queries that are concurrently executed or scheduled within a reasonably short time frame. Olston et al. [54] propose a couple of optimization opportunities for concurrent work sharing. Although shared scans reduce the I/O load, overly aggressive sharing can lead to performance degradations in some cases. References [4] and [53] study scheduling and sharing strategies for scans of large files in MapReduce. The techniques are similar to multi-query optimization known from relational database system research [39, 59]. None of the proposed techniques has been incorporated into one of the presented languages yet.

2.4 Optimization of Plain MapReduce Programs

This section covers approaches that optimize the execution of plain MapReduce jobs. Prior to its execution, a MapReduce job is intercepted and analyzed. Based on that analysis, it is executed in an optimized fashion. In the following, we present two research frameworks that transparently analyze and optimize the execution of Hadoop jobs.

2.4.1 Starfish

Starfish [42, 43] generates optimized parameter settings for arbitrary Hadoop MapReduce jobs. Hadoop provides more than 190 parameters of which about 25 have significant impact on the performance of a job [13]. Manual tuning of these parameters is very difficult since the optimal configuration depends on the job, the input data and the available compute resources. In addition, some parameters interact with others. Therefore, automatic parameter tuning is highly desirable. Starfish was designed as a general self-tuning system for big data analytics [43]. That includes tuning on different abstraction levels, namely job, workflow and workload level. This chapter focuses on job level optimization, i.e., the optimization of a single MapReduce job.

Starfish employs a job profiler, a What-If engine and a cost-based optimizer to improve the execution of arbitrary MapReduce jobs. Prior to optimization, the profiler monitors the job's execution to derive its performance characteristics. A so-called *job profile* is handed to the What-If engine which includes a cost model

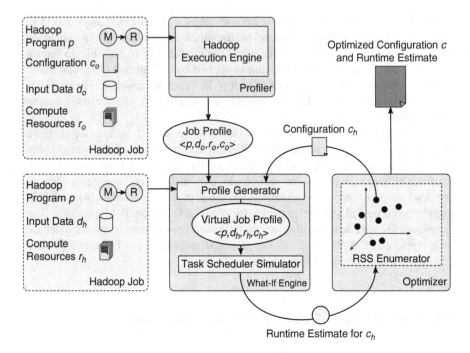

Fig. 2.4 The Starfish architecture

for Hadoop's execution model [41]. The engine simulates the execution of the job for varying environmental conditions, such as different input sizes, compute resources and parameter settings. The cost-based optimizer explores the search space of parameter settings and uses the What-If engine to retrieve cost estimates for given configurations. Figure 2.4 shows how Starfish's components interact with each other. In the following, each of them is discussed in more detail.

2.4.1.1 Profiler

The profiler collects statistical summaries that represent the performance characteristics of a given MapReduce job by monitoring its execution. The Hadoop MapReduce execution pipeline is split up into 13 phases [42]. During profiling, the job's behavior in each phase is monitored. The monitored data is aggregated to derive the characteristic of each phase. The profiler uses multi-level hierarchical aggregation to eagerly reduce the amount of processed profiling data. Finally, the profiler emits a job profile consisting of more than 60 monitored values.

The profiler uses a technique, called *dynamic instrumentation*, to monitor the Hadoop execution framework. During monitoring, solely the Hadoop engine is accessed, not the user-code. Therefore, the approach is not limited to MapReduce jobs written in a specific programming language. Dynamic instrumentation does not cause overhead when profiling is deactivated. The profiler supports two modes of task sampling. First, ad-hoc sampling, where only a fraction of the job's tasks is executed and monitored. In the second mode, the profiler hooks up to a regular run of the job and monitors a fraction of the job's tasks.

2.4.1.2 What-If Engine

Starfish's What-If engine simulates the execution of a hypothetical MapReduce job in two steps. First, a *virtual job profile* is generated based on an actual job profile which was generated by the profiler. Second, the engine simulates the execution of the hypothetical job using the virtual job profile. Both steps are described in more detail in the following.

A job is defined by a MapReduce program p, input data d, resources r and a configuration c. A virtual job profile captures the performance characteristics of a hypothetical job. Given a hypothetical job $< p, d_h, r_h, c_h >$ and an observed job profile for the program p with input data d_o, resources r_o and configuration c_o, the What-If engine computes a virtual profile. The engine captures the impact of the varying configurations c_h and c_o, using a set of mathematical and simulation white-box models which have been published in a technical report [41]. Differences in the input data sets d_h and d_o, and resources r_h and r_o, are taken into account based on dataflow proportionality and cluster homogeneity assumptions. The resulting virtual job profile contains dataflow and cost estimations for the given hypothetical job.

After the virtual job profile has been derived, the What-If engine uses a *Task Scheduler Simulator* to simulate the execution of the hypothetical job based on the virtual job profile. The simulator resembles Hadoop's FIFO scheduler and computes various metrics, such as execution time and amount of local I/O. Based on the derived information, the execution of a hypothetical job can even be visualized. Finally, the computed metrics are returned to the optimizer that called the What-If engine. Experiments show a high precision of the What-If engine's estimations [42].

2.4.1.3 Cost-Based Optimizer

The goal of the cost-based optimizer is to find an optimal parameter configuration for a job that is executed on a specified data set and Hadoop setup. In order to derive that configuration, the optimizer has to explore a high-dimensional, nonlinear, non-convex and multimodal search space [42]. Babu [13] claims that about 25 parameters have significant impact on the performance of a Hadoop job. These parameters are from different domains, such as binary, integer and decimal values, and have varying validity ranges. In addition, some parameters interact with each other. Hence, finding the optimal parameter configuration is a hard problem.

Starfish's optimizer follows two approaches to tackle the optimization problem [42]. First, the highly dimensional search space is split into multiple subspaces with fewer dimensions. Many parameters have only impact on a subsection of the whole MapReduce execution pipeline, such as the Map or the Reduce phase. Parameters that interact or affect the same subsections of the pipeline are clustered. The optimizer processes all clusters independently from each other. Second, the optimizer employs a technique called *Recursive Random Search* (RRS) to find good parameter settings within a cluster. RRS picks arbitrary points in the search space. For each point, the What-If engine is called and computes a cost estimate. In consecutive iterations, the optimizer inspects promising areas in more detail by taking more samples from them. Finally, the optimizer returns the configuration with the shortest estimated runtime found so far. Reference [42] evaluates the quality of the optimizer's decisions.

2.4.2 Manimal

Manimal is a hybrid of a relational database and a MapReduce system [22]. It follows a different optimization approach than Starfish. Manimal employs static code analysis to infer the semantic of arbitrary user code. Based on the findings, the user code is fully transparently modified and executed on the Hadoop framework. The rationale behind Manimal's approach is that certain programming patterns are commonly used in many MapReduce jobs to perform relational operations, such as selection or projection. Manimal aims to identify such patterns and reason about their semantics which is required for optimization.

The system consists of four components. An analyzer performs code analysis and detects optimization opportunities. A rule-based optimizer accesses a catalog that holds information on available indexes and chooses the most promising optimization among all possible optimizations. Finally, the job is executed on a modified Hadoop execution framework called *Execution Fabric*. In the following, we shortly discuss each of the four components focusing on the Analyzer.

2.4.2.1 Analyzer

The Manimal system accepts Hadoop MapReduce jobs written in Java. The analyzer performs static code analysis on the compiled Java byte-code and identifies so-called *optimization descriptors*. An optimization descriptor defines a certain optimization opportunity. At the current state, Manimal supports optimizations of selections and projections, and is able to add compression, i.e., delta compression for numerical values and operations on compressed data [46]. Due to the hidden semantics in user code, the analyzer follows a very conservative policy and only generates an optimization descriptor if it can be undoubtedly identified. Most of the currently supported optimizations are based on indexes. Along with optimization descriptors that require an index, the analyzer emits an index creation program which creates the necessary index from the original job input data.

Manimal's optimizations are well-known from the context of relational database systems. Its contribution comes from the fact that these optimizations are applied to arbitrary user code instead of semantically rich declarative queries. However, the applicability of static code analysis is limited. Map and Reduce implementations of higher-level languages, such as Pig and Jaql, are essentially query interpreters that are customized via runtime program parameters. Therefore, the semantic of the program depends solely on the runtime parameters and cannot be determined by Manimal's analyzer. Also, specialized data structures as used by libraries, such as Mahout [9], pose a big challenge for static code analysis. However, experiments with third-party user code [56] show that Manimal's analyzer is able to identify most optimization opportunities in relational-style jobs [46]. Currently, the system is restricted to single MapReduce jobs. Workflows consisting of multiple jobs can not be optimized as a whole.

2.4.2.2 Optimizer, Catalog and Execution Fabric

Manimal's optimizer applies simple rule-based heuristics [22]. It inspects all optimization descriptors provided by the analyzer and tries to match these with available indexes retrieved from the catalog. Based on that information, the optimizer chooses the index that promises the best speed-up. At the current state, only one optimization conflict might occur. The optimizer solves that conflict by always preferring selection over compression optimization [46]. Finally, the Execution Fabric executes MapReduce jobs optimized by Manimal. Instead of scanning the whole input file, optimized jobs receive their data from the chosen index. If no optimizations were applied, the job is executed using Hadoop's default strategy.

The decision to create an index is not made by the system. Instead, the user has to choose whether and which indexes to create. An index advisory component, such as the ones available for relational database systems [3], is not part of Manimal. Whenever an index is created, a corresponding entry is added to the catalog.

2.5 Parallel Data Flow Systems

Much of MapReduce's popularity can be accounted to its simple programming interface. However, the inflexibility of the programming and execution model often impedes efficient data processing. In recent years, several systems have been built that process arbitrary directed acyclic data flows in a massively parallel fashion. These systems trade ease of task definition for flexibility and performance. However, abstractions, such as query languages and programming models, have been built to ease the definition of data processing tasks. Optimizers compile higher-level task definitions into data flows that are executed in parallel.

This section presents three parallel data flow systems that have varying features and differ in programming interfaces and optimization capabilities.

2.5.1 The Dryad Ecosystem

Dryad is a parallel data processor designed by Microsoft Research. The system processes acyclic data flows at large scales and is a very flexible execution platform for massive parallel analysis tasks. Within Microsoft Research, a couple of projects evolved that target Dryad as execution platform. Among them are DryadLINQ and SCOPE. DryadLINQ extends LINQ for data parallel tasks and provides an optimizer that compiles LINQ programs to Dryad data flows. SCOPE is a SQL-like higher-level language that is also executed on Dryad systems. We briefly introduce the relevant concepts of Dryad and continue discussing the features and optimization techniques of DryadLINQ and SCOPE.

2.5.1.1 Dryad

Dryad is a massively parallel execution engine for data flows [29]. It is designed to run on shared-nothing clusters of hundreds to thousands of commodity servers. The system uses data parallelism to execute data flows in parallel and relieves the programmer from many cumbersome issues of parallel data processing, such as task scheduling, data transport and error handling [45].

Dryad data flows are defined as directed acyclic graphs (DAGs), where vertices run arbitrary user code and edges represent communication channels between vertices. A data flow may have multiple data sources and sinks. The data flows from sources to sinks and passes on its way vertices that run sequential blocks of user code. Dryad processes arbitrary black-box data items. The user code is responsible for serialization and interpretation of the data.

When a Dryad program is executed, the system constructs a communication graph by spanning vertices and channels (see Fig. 2.5). Multiple instances of a task vertex are spawned on multiple machines. Each instance of a vertex processes a subset of the task vertex' input data. In order to establish communication and data transport between vertex instances, communication channels must be replicated as well. Thereby, channels can be constructed in multiple ways, i.e., 1:1, 1:N, and M:N connection patterns, leading to different job semantics. Dryad features three kinds of communication channels: files, TCP pipes, and shared-memory FIFOs. Although, Dryad's programming abstraction uses pipeline semantics, often batch processing is internally used for higher efficiency.

Compared to MapReduce, Dryad offers a more flexible and efficient execution engine. Both use a distributed file systems as default storage system. However, writing well-performing data flow graphs is significantly harder than writing MapReduce jobs. To perform a memory-safe, external join, a Dryad vertex must implement a join strategy, such as an external sort-merge join or a hybrid-hash join, which is a major undertaking. Dryad was specifically designed to be an execution engine for job specifications with a higher abstraction [45].

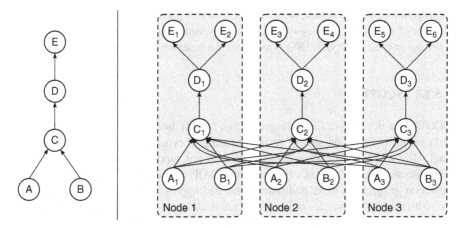

Fig. 2.5 A Dryad program DAG and a Dryad communication graph

2.5.1.2 DryadLINQ

Microsoft's LINQ (Language INtegrated Query) is an approach to tightly integrate a data processing and query framework into the higher-level programming languages of the .NET framework, such as C#, F# or VisualBasic. Queries are stated in the same language in which the application is written using a special LINQ library. In contrast, SQL is an individual language whose queries are often included into application code as strings. Together, LINQ and the hosting programming language, offer characteristics of declarative and imperative languages. Embedded LINQ queries are compiled by so-called *LINQ providers* and issued against data processing systems, such as relational DBMSs or XML processors at runtime.

DryadLINQ provides LINQ extensions for data parallel processing and a LINQ provider for the Dryad execution framework [65]. The tight integration of LINQ into the hosting language allows to state data analysis tasks as a mixture of parallelizable and sequential code sections. DryadLINQ's data model is based on .NET objects, which allows seamless integration with the application code.

The DryadLINQ provider extracts data parallel LINQ code from the application and compiles it into Dryad DAGs. DryadLINQ's optimizer shares many features with traditional optimizers of parallel relational DBMSs [44]. At the current state, it is based on greedy heuristics and applies optimizations, such as pipelining of operations, minimizing repartitioning steps, eager aggregations, and reduction of I/O by using TCP-pipe and memory-FIFO channels. The optimizer derives much semantic from the program code by static typing, static code analysis, and reflection. Using these techniques, it reasons about prevalent data properties, such as partitioning and sorting. Partitioned data stores are supported as well. DryadLINQ features a rich set of user annotations to hint potential optimizations and expected memory consumption of user code. A notable feature is dynamic optimization. DryadLINQ adapts execution plans at runtime in order to perform network-aware aggregations and tree broadcasting, and determines partitioning properties, such as number of partitions and key-ranges for range-partitioning, to handle skewed data. The source code of DryadLINQ is available for research upon request [30].

2.5.1.3   SCOPE

SCOPE is a declarative query language that heavily borrows from SQL [24]. The data model is based on sets of rows consisting of typed attributes. In contrast to Jaql, Hive and Pig, SCOPE's data model does not support nested data structures. The operator set and syntax originates from SQL as well. The language supports selection, projection, inner and outer joins, and aggregations. SCOPE queries can be written as nested queries or in an imperative style, similar to Pig and Jaql, where multiple individual statements are linked. Nested subqueries are not supported. Similar to Pig, Hive and Jaql, SCOPE is extensible and allows to define three types of parallelizable UDFs, called *Process*, *Reduce* and *Combine*, which correspond to the first-order functions of the MapReduce programming model.[4]

SCOPE's optimizer compiles queries to parallel Dryad data flows. It is based on the Cascades framework [36], uses transformations and features a cost model. The optimizer includes transformations for selection and projection push-down, eager aggregation, and chooses appropriate partitioning schemes. A special focus was put on optimizing partitioning, sorting and grouping [66]. These data properties are crucial for data parallel processing, required for many operations, and very expensive to establish. Minimizing the number of partitioning, sorting and grouping steps is a major optimization goal for parallel environments. When reasoning about existing data properties, the SCOPE optimizer takes functional dependencies and constraints on the data into account. The rather simple data model relieves such reasoning. The order of joins is chosen based on heuristics, such as the preference of equi-joins.

SCOPE uses many execution strategies known from parallel database systems, such as alternative join algorithms and sort-based or hash-based grouping. The system features basic runtime adaptions, such as topology-aware tree aggregations.

2.5.2   Asterix

The Asterix project researches and builds a scalable information management system that operates on large clusters of commodity servers [11]. Asterix is a joint project of UC Irvine, UC Riverside and UC San Diego. The goal of the project is to store, index, process, analyze and query semi-structured data [16]. Key features of the system are support for short and long running queries over structured and semi-structured evolving data sets.

---

[4]Process is equivalent to Map.

The Asterix software stack consists of the Hyracks execution framework, index structures, the Asterix query language (AQL) and algebra (Algebrix), and a query optimizer. Queries stated in AQL are compiled by the optimizer to Hyracks jobs and executed exploiting the available indexes. While the Hyracks framework is publicly available, the remaining components of the system are planned to be released as open-source in the future.

In the following, we highlight the features of Hyracks, describe the Asterix data model and query language, and discuss the Asterix optimizer.

2.5.2.1 Hyracks

Hyracks is a data parallel execution framework for data-intensive compute tasks, which operates on large shared-nothing clusters [20]. Similar to Dryad, Hyracks was specifically designed to serve as an execution platform for higher-level languages and is based on well-known techniques from parallel database systems.

Hyracks jobs are specified as DAGs with nodes being Hyracks operators and edges being connectors. An important design issue was extensibility. Although Hyracks comes with a set of common operators and connectors, own operators and connectors can be defined. The system features an API to specify characteristics and requirements of operators, such as memory consumption for user-defined operators. This information is used to improve resource management and scheduling. The set of included operators and connectors includes Mappers, Joiners, Aggregators, Partitioners and Replicators. In contrast to Dryad vertices, Hyracks operators are not atomic. Instead they are composed of activities. Hyracks plans explicitly specify dependencies among activities. For example, the hash-join operator consists of a hash-table build activity and a probe activity that cannot be started before the building activity is finished. Hyracks' data model is a generalization of MapReduce's key/value pair data model and is based on tuples with fields of arbitrary types. Furthermore, the system supports many operations on binary data.

Hyracks fragments jobs along activity dependencies into fully pipelined segments, called *stages*. When executing a job, Hyracks schedules and executes a stage as soon as its dependencies are fulfilled. The lazy scheduling allows the system to use the latest available information on resources and data characteristics.

2.5.2.2 Asterix Data Model and Query Language

The *Asterix Data Model* (ADM) is based on concepts from JSON, Avro and structured data types of object database systems [16]. ADM is a nested data model of primitive and derived types. Derived types are composed of primitive or derived types. Supported derived types are enums, records, ordered and unordered lists and unions. Also, optional values can be specified. Similar to Jaql, ADM features open schemas, i.e., data items are allowed to contain more information than specified in the schema. However, schema enables optimizations and efficient processing. ADM organizes data in data sets that can be indexed, partitioned and replicated.

AQL is inspired by XQuery, focusing on relational and nesting operations [16]. XQuery's document-related and XML specific features are omitted. Nested subqueries are supported. AQL includes data set manipulation operations, such as inserts and updates. Updates are performed using a timestamp model.

2.5.2.3 Asterix Algebra and Optimizer

Asterix features an algebra, called *Algebrix*. Prior to optimization, AQL queries are parsed, checked and transformed into an algebra tree on which optimizations are performed. Among the presented systems, Asterix is the only approach that separates language from algebra. This design decision eases the porting of other higher-level languages, such as Pig, Jaql or Hive, to the Asterix system. Algebrix supports operations, such as selections, projections, joins, aggregations, group-by and nesting [16]. In addition, data-model specific computations, such as filter predicates and new data values, are included in the algebra. At the current state, user-defined data types are not supported.

Static optimizations based on heuristics are preformed on the algebra tree [16]. The optimizer adds index accesses and aims to minimize data re-organizations, such as partitioning, sorting and grouping. The resulting job is stage-wise executed by Hyracks. A stage is executed as soon as its prerequisites are fulfilled. Thereby, Hyracks decides on the degree of parallelism, node scheduling and resource allocation. A distinguishing feature of Asterix is deferred optimization. The optimizer can insert a so-called *choose-plan operator* instead of a physical operator, cf. [38]. Prior to execution, the choose-plan operator decides on the actual physical operator to use based on information gathered while executing preceding stages. However, this approach does not allow to change the global execution plan or to re-optimize at runtime.

2.5.3 Stratosphere

The Stratosphere system [61] is an open-source research prototype resulting from the Stratosphere project jointly researched by TU Berlin, HU Berlin and HPI Potsdam. At the current state, the system consists of three components: the parallel programming model PACT, the flexible and massively parallel data flow processor Nephele and an optimizer that compiles PACT programs into Nephele data flows. In the following, all three components are shortly discussed.

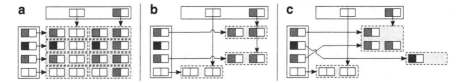

Fig. 2.6 Input contracts: (**a**) Cross, (**b**) Match and (**c**) CoGroup

2.5.3.1 PACT Programming Model

The PACT programming model is based on so-called *Parallelization Contracts* (PACTs). PACT is a generalization and extension of the MapReduce programming model [15]. Both have a couple of concepts in common, such as the key/value pair data model and the concept of second-order system, and first-order user functions. Second-order functions are called *Input Contracts* in the context of the PACT programming model. In fact, PACT is a superset of MapReduce. Hence, all MapReduce jobs can be expressed with PACT.[5]

However, PACT differs in three main aspects from MapReduce. First, it includes additional Input Contracts that complement Map and Reduce. Cross, Match and CoGroup associate key/value pairs from two inputs into subsets that are independently processed by the first-order user functions (see Fig. 2.6). Cross builds a Cartesian product and calls the user code for each element of the cross product. Match basically performs an equi-join on the key and calls its user function for each pair of key/value pairs that share the same key. CoGroup is a hybrid of Match and Reduce. Second, the PACT programming model features user code annotations called Output Contracts that exhibit information about the user code behavior. This information is used by the PACT optimizer to generate efficient execution plans. Finally, PACT programs are composed as arbitrary DAGs with possibly multiple data sources and sinks, i.e., they do not have a static structure. Vertices of a DAG are PACTs, which consist of an Input Contract, user code and an optional Output Contract.

Due to the higher expressiveness resulting from the richer set of second-order functions and the flexible program structure, the PACT programming model is better suited to define complex data analysis tasks than MapReduce. Alexandrov et al. [6] provide a detailed comparison of MapReduce and the PACT programming model.

---

[5]This is true for the pure programming model, not necessarily for its implementations, such as Hadoop.

2.5.3.2 Nephele Data Flow Engine

Nephele is an efficient parallel execution engine for cloud environments [64]. The system processes data flows specified as DAGs. From the data flow specification point of view, Nephele is similar to Dryad. Data flows are composed from vertices and communication channels. Vertices run arbitrary user code and channels transport data from one vertex to another. Vertices can be annotated with explicit scheduling information, such as intra-node parallelism or task instance co-location.

Nephele was designed to run in cloud environments. It connects to infrastructure-as-a-service (IAAS) providers, such as Amazon EC2 or the Eucalyptus framework. Nephele is able to adapt compute resources to a task's requirements, by dynamically starting and stopping virtual machines while the task is executed.

2.5.3.3 PACT Optimizer

The optimizer compiles PACT programs into Nephele data flows. The optimization potential originates from two properties of the programming model, the declarative character of Input Contracts and the Output Contract code annotations [15].

Input Contracts declaratively specify the partitioning of data into subsets that are processed by independent user function calls. The choice of the strategy to establish the partitioning is a degree of freedom for the system. For example, a Match performs an equi-join on the key and can be processed using a repartition or broadcast shipping strategy, and a sort-merge or hash-based local strategy. Optional Output Contracts reveal information about behavior of user code that is relevant for the optimizer. When executing a PACT program, data is frequently partitioned, sorted or grouped, before the user code is called. Since user functions can perform arbitrary operations, the system can not infer that those properties are also valid for the function's output. Output Contracts help the system to infer which data properties are existent at a specific point in the data flow graph.

The Stratosphere system uses a cost-based optimizer to compile PACT programs into Nephele DAGs [5, 15]. At the current state, the optimizer aims to reduce network communication cost. Input sizes and cardinalities are obtained by light-weight sampling. Size estimates for intermediate data sets are computed based on the sampled information, available Output Contracts and compiler hints. During optimization, candidate execution plans are enumerated and costed. The search space is limited by the relatively few execution strategies which are available at the moment. Therefore, the full search space can be explored. The optimizer tracks and tries to reuse existing data properties such as partitioning, sorting and grouping. This is very similar to the approach that SCOPE's optimizer follows [66]. Due to black-box data types and user functions, logical rewrites cannot be applied. PACT's optimization process resembles physical optimization in parallel RDBMSs. However, PACT optimizes data flows with arbitrary user code instead of algebraic queries.

In order to improve PACT's optimization capabilities, it is planed to semantically enrich the programming model. Promising approaches are a generalized tuple data model, similar to Hyracks' data model, and additional user code annotations that allow to structurally modify PACT programs.

2.6 Comparing State-of-the-Art Systems

Query optimizers aim to derive the most efficient execution plan for a given query, based on the query semantics, available execution strategies, and additional information such as data statistics and system state. The systems presented in this chapter differ in their prerequisites for optimization. Figure 2.7 shows the architectural stacks of all presented systems and highlights their optimizers. In this section, we compare the presented approaches and discuss how design decisions affect their optimization capabilities.

2.6.1 Programming Interface

Traditional query optimization relies on full or partial knowledge about the semantics of the query to be optimized. Declarative languages, such as SQL, provide that information by specifying the result rather than giving instructions to follow.[6]

Higher-level languages, such as Pig, Hive, SCOPE and AQL, follow a similar approach. Although some feature a rather imperative style, all exhibit declarative elements, such as join operators. In order to support extensibility and improve applicability, all presented languages feature the definition and integration of custom operators, user-defined functions or hand-written MapReduce jobs. Jaql differs from all languages due to its physical transparency feature that allows users to specify tasks on varying levels of abstraction, i.e., using high-level declarative operators and low-level physical operators within the same script.

In contrast to higher-level languages, DryadLINQ does not feature an own language. Instead, it is tightly integrated into hosting languages. This approach is very appealing due to its ease of use. The DryadLINQ provider parses the relevant code fragments and derives the semantic information required for optimization.

Other programming abstractions are parallel programming models, such as MapReduce and PACT. Manimal and Starfish use completely different techniques to optimize the execution of MapReduce programs. While Manimal employs static code analysis to derive certain semantic aspects of a job, such as selection or projection, Starfish does not use semantic information. Instead, Starfish profiles the runtime behavior of the job to generate optimized parameter settings.

---

[6]User-defined functions (UDFs) incorporate semantics a query optimizer cannot reason about.

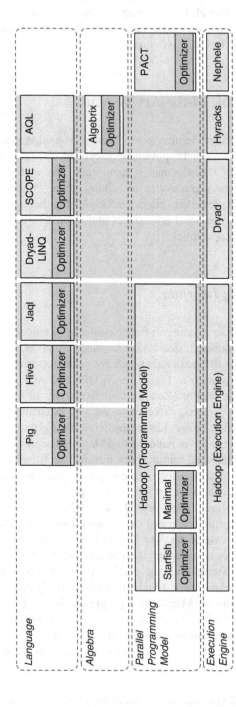

Fig. 2.7 Comparing massively parallel data flow system stacks

PACT is a generalization and extension of the MapReduce programming model. Input Contracts declaratively specify the parallelization of user-defined functions. Optional semantic user-code annotations provide additional information used for optimization.

While approaches based on parallel programming models must cope with arbitrary user-code, higher-level languages exhibit detailed information about the semantics of an analysis task. However, extensive usage of the extensibility features of higher-level languages can diminish that advantage.

2.6.2 Execution Platform

We presented four execution platforms for parallel data analysis tasks. The most prominent is Hadoop which implements the MapReduce execution model. As discussed in Sect. 2.2.2, the structure of the MapReduce execution model is static. Hadoop provides a number of configuration parameters [13] and interfaces for user-defined code [28] that can control the execution to a certain degree. For example, hash-partitioning can be altered to range-partitioning. However, the partitioning step cannot be omitted. Due to its conservative approach to ensure fault-tolerance, the MapReduce programming model does not feature pipelined execution. In order to process complex tasks, multiple MapReduce jobs can be chained by writing and reading temporary files from a distributed file system. These limitations lead to fewer degrees of freedom when choosing efficient physical execution plans.

Dryad, Hyracks and Nephele execute arbitrary acyclic data flows. In contrast to the MapReduce execution model, these systems provide a much more flexible and lower-level interface to specify parallel data flows. Due to their flexibility, these frameworks can be used to implement many data processing techniques known from parallel relational database systems. Although they offer very similar programming interfaces, Dryad, Hyracks and Nephele differ in some details. Dryad offers hooks to change execution plans at runtime, which can be used for continuous optimization. Hyracks features deferred stage-wise scheduling to take the latest available information on the system into account. Nephele was specifically designed for cloud environments and is able to start and stop compute nodes during execution.

2.6.3 Optimization

Recent trends in information management, such as MapReduce and the NoSQL movement, avoid the requirements for schema information and time-consuming data imports. Instead, these systems aim to provide ad-hoc data processing of varying data sets. Most of the presented systems support the processing of semi-structured data models or arbitrary data types. However, schema information and data statistics are crucial for certain optimizations. Table 2.1 lists and categorizes the systems that were discussed in this chapter.

Table 2.1 A comparison of the discussed approaches for optimization of massively parallel data flows

System	Programming interface	Data model	Optimization	Catalog
Higher-level languages for MapReduce				
Pig	Higher-level language	Semi-structured, nested	Rule-based	No
Hive	Higher-level language	Semi-Structured, nested	Rule-based	Yes
Jaql	Higher-level language with physical transparency	Semi-structured, nested, partial	Rule-based	No
General MR optimization				
Starfish	MapReduce PM	Key/value pairs	Cost-based	No
Manimal	MapReduce PM	Key/value pairs	Rule-based	Yes
Parallel data flow systems				
DryadLINQ	Higher-level language	.NET objects	Rule-based	Yes
SCOPE	Higher-level language	Tuples	Rule- & cost-based	No
Asterix	Higher-level language	Semi-structured, nested, open	Rule-based	Yes
Stratosphere	PACT PM	Key/value pairs	Cost-based	No

Some presented systems, such as Pig, Hive, Jaql and DryadLINQ, perform rule-based optimization by applying logical transformations. Many rewrite rules, such as selection and projection push-down, which are known from relational database optimizers. Subsequent to logical optimization, the query is compiled into a physical execution plan. The physical optimization of Hive, Jaql and Pig is also based on heuristics. Due to the lack of data statistics, most systems are not able to perform cost-based optimizations. To overcome this limitation, Pig, Jaql and Hive feature compiler hints and let the users to decide the physical strategies to execute join operations. Jaql's physical transparency allows users to directly use physical operators.

SCOPE, PACT and Starfish feature cost-based optimizers. Starfish does not optimize the MapReduce program itself. Instead it optimizes the execution of the program by choosing well-performing parameter settings. SCOPE's optimizer is very close to traditional database optimizers. Due to the higher-level language interface, the semantics of a query are well-understood by the optimizer. The optimization process starts by applying logical rewrite rules. In addition, metadata, such as schema, partitioning and statistics, can be taken into account for cost-based optimization. In contrast, PACT does not perform logical information due to lack of task semantics, which are hidden within black-box user-code. However, cost-based optimization is performed to derive an efficient physical execution data flow. In order to acquire information about the input data, PACT's optimizer draws small samples from the input files. Additional statistical hints, such as selectivity and record width annotated by the programmer, improve the optimizer's estimates.

Hive, Manimal, Asterix and DryadLINQ feature catalogs to store metadata. Hive's Metastore provides schema and partitioning information to the optimizer. In the case of selections on partitioned attributes, the optimizer restricts scans to necessary partitions. Manimal and Asterix use their catalogs to store information on available indexes, which can be used to significantly improve query runtimes. In addition, Asterix's catalog includes schema and partitioning information similar to Hive. DryadLINQ and Asterix feature partitioned storage, such that initial partitioning steps can be omitted.

2.7 Future Research and Conclusion

The preceding sections presented the state of the art in optimization of parallel data flows. This section concludes the chapter and proposes future areas of research.

Massively parallel data analytics is an ongoing research topic, with many research projects and researchers working in this domain. The size of data sets and the complexity of data analysis tasks will continue to grow at high rates. Optimization is crucial to achieve high performance, when performing sophisticated analysis on huge amounts of data. In this chapter, we presented and compared different approaches to optimize parallel data flows.

All presented approaches are limited by missing information on the semantics of tasks, the data to process, and the available resources. Some systems statically choose physical execution plans or delegate important decisions to the user. Others perform cost-based optimization with limited statistical information that causes estimation errors. Both approaches often lead to wrong decisions, which can have significant impact on the performance of a task.

Robust optimization has been researched in the context of relational database systems and aims to reduce the impact of cardinality estimation errors [1, 12, 14]. Instead of choosing the most promising execution plan with minimal estimated costs, the optimizer decides to choose a possibly more expensive plan that exhibits a more robust behavior with respect to estimation errors. Robust behavior in this context means that an execution strategy does not cause significantly higher runtime costs if the assumptions of the optimizer do not hold. A robust optimizer must be able to reason about the quality of its estimates, in order to pick the best performing strategy in the presence of reliable estimates and a robust strategy in case of uncertain information. Robust optimization is a promising technique to cope with the lack of information about the input data, in the context of massively parallel analytics.

Further approaches with good prospects are re-optimization at runtime and adaptive execution frameworks. Some work on re-optimization of query execution plans and adaptive algorithms in relational database systems has been performed [37, 49]. However, adapting and modifying execution plans is much more difficult in parallel setups than in centralized environments, since global changes must be synchronized. Systems that incorporate such approaches can react on estimation errors and unforeseeable changes in the execution environment. These are desirable features, especially for long running tasks. DryadLINQ's and SCOPE's runtime adaptions, and Hyracks' stage-wise query execution are only first steps in this direction.

References

1. Abhirama, M., Bhaumik, S., Dey, A., Shrimal, H., Haritsa, J.R.: On the stability of plan costs and the costs of plan stability. PVLDB **3**(1), 1137–1148 (2010)
2. Afrati, F.N., Ullman, J.D.: Optimizing joins in a map-reduce environment. In: EDBT, Lausanne, pp. 99–110 (2010)
3. Agrawal, S., Chaudhuri, S., Narasayya, V.R.: Automated selection of materialized views and indexes in SQL databases. In: VLDB, Cairo, pp. 496–505 (2000)
4. Agrawal, P., Kifer, D., Olston, C.: Scheduling shared scans of large data files. PVLDB **1**(1), 958–969 (2008)
5. Alexandrov, A., Battré, D., Ewen, S., Heimel, M., Hueske, F., Kao, O., Markl, V., Nijkamp, E., Warneke, D.: Massively parallel data analysis with pacts on nephele. PVLDB **3**(2), 1625–1628 (2010)
6. Alexandrov, A., Ewen, S., Heimel, M., Hueske, F., Kao, O., Markl, V., Nijkamp, E., Warneke, D.: Mapreduce and pact – comparing data parallel programming models. In: BTW, Kaiserslautern, pp. 25–44 (2011)
7. Apache Hadoop: http://hadoop.apache.org
8. Apache Hive: http://hive.apache.org
9. Apache Mahout: http://mahout.apache.org
10. Apache PIG: http://pig.apache.org
11. Asterix: A highly scalable parallel platform for semi-structured data management and analysis. http://asterix.ics.uci.edu
12. Babcock, B., Chaudhuri, S.: Towards a robust query optimizer: a principled and practical approach. In: SIGMOD conference, Baltimore, pp. 119–130 (2005)
13. Babu, S.: Towards automatic optimization of mapreduce programs. In: SoCC, Indianapolis, pp. 137–142 (2010)
14. Babu, S., Bizarro, P., DeWitt, D.J.: Proactive re-optimization with rio. In: SIGMOD conference, Baltimore, pp. 936–938 (2005)
15. Battré, D., Ewen, S., Hueske, F., Kao, O., Markl, V., Warneke, D.: Nephele/PACTs: a programming model and execution framework for web-scale analytical processing. In: SoCC'10: Proceedings of the ACM Symposium on Cloud Computing, Indianapolis, pp. 119–130. ACM, New York (2010)
16. Behm, A., Borkar, V.R., Carey, M.J., Grover, R., Li, C., Onose, N., Vernica, R., Deutsch, A., Papakonstantinou, Y., Tsotras, V.J.: Asterix: towards a scalable, semistructured data platform for evolving-world models. Distrib. Parallel Databases **29**(3), 185–216 (2011)
17. Bernstein, P.A., Goodman, N., Wong, E., Reeve, C.L., Jr., J.B.R.: Query processing in a system for distributed databases (SDD-1). ACM Trans. Database Syst. **6**(4), 602–625 (1981)
18. Beyer, K., Ercegovac, V., Gemulla, R., Balmin, A., Eltabakh, M., Kanne, C.C., Ozcan, F., Shekita, E.J.: Jaql: a scripting language for large scale semistructured data analysis. PVLDB **4**, 1272–1283 (2011)
19. Blanas, S., Patel, J.M., Ercegovac, V., Rao, J., Shekita, E.J., Tian, Y.: A comparison of join algorithms for log processing in mapreduce. In: SIGMOD conference, Indianapolis, pp. 975–986 (2010)
20. Borkar, V.R., Carey, M.J., Grover, R., Onose, N., Vernica, R.: Hyracks: a flexible and extensible foundation for data-intensive computing. In: ICDE, Hannover, pp. 1151–1162 (2011)
21. Bryant, R.E.: Data-intensive supercomputing: the case for disc. Tech. Rep. CMU-CS-07-128, School of Computer Science, Carnegie Mellon University (2007)
22. Cafarella, M.J., Ré, C.: Manimal: relational optimization for data-intensive programs. In: WebDB, Indianapolis (2010)
23. Cascading: http://www.cascading.org/
24. Chaiken, R., Jenkins, B., Larson, P.Å., Ramsey, B., Shakib, D., Weaver, S., Zhou, J.: SCOPE: easy and efficient parallel processing of massive data sets. PVLDB **1**(2), 1265–1276 (2008)

25. Cohen, J., Dolan, B., Dunlap, M., Hellerstein, J.M., Welton, C.: Mad skills: new analysis practices for big data. PVLDB **2**(2), 1481–1492 (2009)
26. Dean, J., Ghemawat, S.: MapReduce: simplified data processing on large clusters. In: OSDI, San Francisco, pp. 137–150 (2004)
27. DeWitt, D.J., Gray, J.: Parallel database systems: the future of high performance database systems. Commun. ACM **35**(6), 85–98 (1992)
28. Dittrich, J., Quiané-Ruiz, J.A., Jindal, A., Kargin, Y., Setty, V., Schad, J.: Hadoop++: making a yellow elephant run like a cheetah (without it even noticing). PVLDB **3**(1), 518–529 (2010)
29. Dryad – Microsoft Research: http://research.microsoft.com/projects/Dryad
30. DryadLINQ – Microsoft Research: http://research.microsoft.com/projects/DryadLINQ
31. Eltabakh, M.Y., Tian, Y., Özcan, F., Gemulla, R., Krettek, A., McPherson, J.: Cohadoop: flexible data placement and its exploitation in hadoop. PVLDB **4**(9), 575–585 (2011)
32. Fender, P., Moerkotte, G.: A new, highly efficient, and easy to implement top-down join enumeration algorithm. In: ICDE, Hannover, pp. 864–875 (2011)
33. Floratou, A., Patel, J.M., Shekita, E.J., Tata, S.: Column-oriented storage techniques for mapreduce. PVLDB **4**(7), 419–429 (2011)
34. Gates, A., Natkovich, O., Chopra, S., Kamath, P., Narayanam, S., Olston, C., Reed, B., Srinivasan, S., Srivastava, U.: Building a highlevel dataflow system on top of mapreduce: the pig experience. PVLDB **2**(2), 1414–1425 (2009)
35. Ghemawat, S., Gobioff, H., Leung, S.T.: The Google file system. In: SOSP, Bolton Landing New York, pp. 29–43 (2003)
36. Graefe, G.: The cascades framework for query optimization. IEEE Data Eng. Bull. **18**(3), 19–29 (1995)
37. Graefe, G.: A generalized join algorithm. In: BTW, Kaiserslautern, pp. 267–286 (2011)
38. Graefe, G., Ward, K.: Dynamic query evaluation plans. In: Proceedings of the 1989 ACM SIGMOD International conference on Management of Data, SIGMOD '89, Portland, pp. 358–366. ACM, New York (1989).
39. Gupta, A., Sudarshan, S., Viswanathan, S.: Query scheduling in multi query optimization. In: IDEAS, Grenoble, pp. 11–19 (2001)
40. Haas, L.M., Freytag, J.C., Lohman, G.M., Pirahesh, H.: Extensible query processing in starburst. In: SIGMOD conference, Portland, pp. 377–388 (1989)
41. Herodotou, H.: Hadoop performance models. Tech. rep., Duke Computer Science (2010). http://www.cs.duke.edu/~hero/files/hadoop-models.pdf
42. Herodotou, H., Babu, S.: Profiling, what-if analysis, and cost-based optimization of mapreduce programs. PVLDB **4**, 1111–1122 (2011)
43. Herodotou, H., Lim, H., Luo, G., Borisov, N., Dong, L., Cetin, F.B., Babu, S.: Starfish: a self-tuning system for big data analytics. In: CIDR, Asilomar, pp. 261–272 (2011)
44. Isard, M., Yu, Y.: Distributed data-parallel computing using a high-level programming language. In: SIGMOD conference, Providence, pp. 987–994 (2009)
45. Isard, M., Budiu, M., Yu, Y., Birrell, A., Fetterly, D.: Dryad: distributed data-parallel programs from sequential building blocks. In: EuroSys, Lisbon, pp. 59–72 (2007)
46. Jahani, E., Cafarella, M.J., Ré, C.: Automatic optimization for mapreduce programs. PVLDB **4**(6), 385–396 (2011)
47. Kossmann, D.: The state of the art in distributed query processing. ACM Comput. Surv. **32**(4), 422–469 (2000)
48. Lin, Y., Agrawal, D., Chen, C., Ooi, B.C., Wu, S.: Llama: leveraging columnar storage for scalable join processing in the mapreduce framework. In: SIGMOD conference, Athens, pp. 961–972 (2011)
49. Markl, V., Raman, V., Simmen, D.E., Lohman, G.M., Pirahesh, H.: Robust query processing through progressive optimization. In: SIGMOD conference, Paris, pp. 659–670 (2004)
50. Mehta, M., DeWitt, D.J.: Data placement in shared-nothing parallel database systems. VLDB J. **6**(1), 53–72 (1997)
51. Moerkotte, G., Neumann, T.: Dynamic programming strikes back. In: SIGMOD conference, Vancouver, pp. 539–552 (2008)

52. Nippl, C., Mitschang, B.: Topaz: a cost-based, rule-driven, multi-phase parallelizer. In: VLDB, New York City, pp. 251–262 (1998)
53. Nykiel, T., Potamias, M., Mishra, C., Kollios, G., Koudas, N.: Mrshare: sharing across multiple queries in mapreduce. PVLDB 3(1), 494–505 (2010)
54. Olston, C., Reed, B., Silberstein, A., Srivastava, U.: Automatic optimization of parallel dataflow programs. In: USENIX Annual Technical Conference, Boston, pp. 267–273 (2008)
55. Olston, C., Reed, B., Srivastava, U., Kumar, R., Tomkins, A.: Pig Latin: a not-so-foreign language for data processing. In: SIGMOD conference, Vancouver pp. 1099–1110 (2008)
56. Pavlo, A., Paulson, E., Rasin, A., Abadi, D.J., DeWitt, D.J., Madden, S., Stonebraker, M.: A comparison of approaches to large-scale data analysis. In: SIGMOD conference, Providence, pp. 165–178 (2009)
57. Pike, R., Dorward, S., Griesemer, R., Quinlan, S.: Interpreting the data: parallel analysis with sawzall. Sci. Program. 13(4), 277–298 (2005)
58. Selinger, P.G., Astrahan, M.M., Chamberlin, D.D., Lorie, R.A., Price, T.G.: Access path selection in a relational database management system. In: SIGMOD conference, Boston, pp. 23–34 (1979)
59. Sellis, T.K.: Multiple-query optimization. ACM Trans. Database Syst. 13(1), 23–52 (1988)
60. Szalay, A., Gray, J.: Science in an exponential world. Nature 440(23), 413–414 (2006)
61. The Stratosphere Project: http://stratosphere.eu
62. Thusoo, A., Sarma, J.S., Jain, N., Shao, Z., Chakka, P., Anthony, S., Liu, H., Wyckoff, P., Murthy, R.: Hive – a warehousing solution over a map-reduce framework. PVLDB 2(2), 1626–1629 (2009)
63. Thusoo, A., Sarma, J.S., Jain, N., Shao, Z., Chakka, P., 0002, N.Z., Anthony, S., Liu, H., Murthy, R.: Hive – a petabyte scale data warehouse using hadoop. In: ICDE, Long Beach, pp. 996–1005 (2010)
64. Warneke, D., Kao, O.: Nephele: efficient parallel data processing in the cloud. In: SC-MTAGS, Portland (2009)
65. Yu, Y., Isard, M., Fetterly, D., Budiu, M., Erlingsson, Ú., Gunda, P.K., Currey, J.: DryadLINQ: a system for general-purpose distributed data-parallel computing using a high-level language. In: OSDI, San Diego, pp. 1–14 (2008)
66. Zhou, J., Larson, P.Å., Chaiken, R.: Incorporating partitioning and parallel plans into the scope optimizer. In: ICDE, Long Beach, pp. 1060–1071 (2010)

Chapter 3
Mining Tera-Scale Graphs with "Pegasus": Algorithms and Discoveries

U Kang and Christos Faloutsos

Abstract How do we find patterns and anomalies, on graphs with billions of nodes and edges, which do not fit in memory? How to use parallelism for such Tera- or Peta-scale graphs? We propose a carefully selected set of fundamental operations, that help answer those questions, including diameter estimation, connected components, and eigenvalues. We package all these operations in PEGASUS, which, to the best of our knowledge, is the first such library, implemented on the top of the Hadoop platform, the open source version of MapReduce.

One of the key observations in this work is that many graph mining operations are essentially repeated matrix-vector multiplications. We describe a very important primitive for PEGASUS, called GIM-V (Generalized Iterative Matrix-Vector multiplication). GIM-V is highly optimized, achieving (a) good scale-up on the number of available machines, (b) linear running time on the number of edges, and (c) more than nine times faster performance over the non-optimized version of GIM-V.

Finally, we run experiments on real graphs. Our experiments run on M45, one of the largest Hadoop clusters available to academia. We report our findings on several real graphs, including one of the largest publicly available Web graphs with 6,7 billion edges. Some of our most impressive findings are (a) the discovery of adult advertisers in the who-follows-whom on Twitter, and (b) the 7-degrees of separation in the Web graph.

U Kang (✉)
Department of Computer Science, KAIST University, 291 Daehak-ro, Yuseong-gu,
Daejeon 305-701, Republic of Korea
e-mail: ukang@cs.kaist.ac.kr

C. Faloutsos
School of Computer Science, Carnegie Mellon University, 5000 Forbes Ave, Pittsburgh,
PA 15213, USA
e-mail: christos@cs.cmu.edu

A. Gkoulalas-Divanis and A. Labbi (eds.), *Large-Scale Data Analytics*,
DOI 10.1007/978-1-4614-9242-9_3, © Springer Science+Business Media New York 2014

3.1 Introduction

Graphs are ubiquitous: computer networks, social networks, mobile call networks, and the World Wide Web, to name a few. Spurred by the lower cost of storage, the success of social networking websites and Web 2.0 applications, and the high availability of data sources, graph data are being generated at unprecedented size. They are now measured in terabytes or even petabytes, with billions of nodes and edges. Historically, however, most graph mining algorithms were designed under the assumption that the graphs would fit in the main memory of a workstation, or a single disk at its largest. The above-mentioned graphs violate these assumptions. They require us to confront our long-held assumption, and to redesign the algorithms so they can work with these new breed of massive graphs. We surveyed promising frameworks that supported parallel computation, on which we could develop such massively-scalable algorithms. We selected Hadoop, an open-source implementation of MapReduce[8], due to its extreme scalability. On the top of Hadoop, we developed the PEGASUS package [16, 30], available at http://www.cs.cmu.edu/~pegasus, an open source library for mining very large graphs.

In this chapter, we first address the research question: what patterns and anomalies can we discover in huge, real-world graphs with billions of nodes and edges? Huge graphs have interesting patterns or regularities, such as those in their connected components, radii, triangles, etc. Discovering these patterns helps us to spot anomalies, a capability that is useful in a wide spectrum of applications, such as cyber-security (computer networks), phone companies (fraud detection) and social networks (spammer detection).

The second question we investigate is how to design efficient algorithms for PEGASUS to handle such massive graphs. There are several challenges. First, can we formulate graph mining algorithms using simple operations that can be efficiently implemented on MapReduce? Second, how to store the graphs efficiently to minimize storage space and to enable fast graph queries? Finally, how to exploit the data distribution (e.g., skewness) for designing faster MapReduce algorithms?

The rest of this chapter is organized as follows. Section 3.2 presents the related work. In Sect. 3.3, we present the discoveries in real world, large scale graphs. Section 3.4 describes the algorithms for large graph mining, which enabled the discoveries in Sect. 3.3. We conclude this chapter in Sect. 3.5.

3.2 Related Work

In this section, we review related work on MapReduce, Hadoop and large scale graph mining with Hadoop.

MapReduce is a programming framework [1, 8] for processing huge amounts of unstructured data in a massively parallel way. MapReduce has two major advantages: (a) the programmer is oblivious of the details of the data distribution, replication, load balancing etc., and (b) the programming concept is familiar, i.e., the concept of functional programming. Briefly, the programmer needs to provide only two functions, a *Map* and a *Reduce*. The typical framework is as follows [24]: (a) the *map* stage sequentially passes over the input file and outputs (key, value) pairs; (b) the *shuffling* stage groups of all values by key, and (c) the *reduce* stage processes the values with the same key and outputs the final result. Hadoop provides the Distributed File System (HDFS) and PIG, a high level language for data analysis [34].

Large scale graph mining has attracted significant interest both from academia and industry: there have been works based on shared memory system [29], and Bulk Synchronous Parallel model [30]. Large graph mining on Hadoop [16, 30] has been used for various large scale graph mining applications due to its power, simplicity, fault tolerance and low maintenance costs.

3.3 Discoveries

In this section, we present the discoveries on large, real world graphs. The discoveries include the patterns and anomalies in radius plots, connected components and triangles. Table 3.1 lists the graphs that were used. The experiments were performed in Yahoo!'s M45 Hadoop cluster, one of the largest Hadoop clusters available to academia with 480 machines, 1.5 petabyte storage and 3.5 TB memory in total.

3.3.1 Radius Plots

What are the central nodes and outliers in graphs? How close are nodes in graphs? These questions are answered by a radius plot, which is the distribution of the radius of nodes. The radius $r(v)$ of node v is the distance between v and a reachable node farthest away from v. The diameter of a graph is the maximum radius over all nodes. The effective radius and the effective diameter are defined as the 90 % percentile of the radius and the diameter, respectively [17, 19]. We analyze the diameter and radii

Table 3.1 Graphs used (M: million. K: thousand)

Name	Nodes	Edges	Description
YahooWeb	1,413M	6,636M	Web links in 2002
Twitter	63M	1,838M	Who follows whom in Nov. 2009
LinkedIn	7.5M	58M	Person-person in 2006
U.S. Patent	6M	16M	Patent citations
Wikipedia	3.5M	42M	Document citations
Random	177K	1,977M	Synthetic Erdős-Rényi graphs

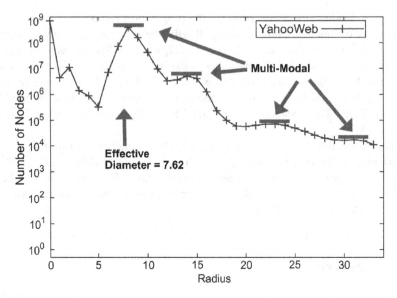

Fig. 3.1 Radius plot of the YahooWeb graph. Notice the effective diameter is surprisingly small. Also notice the multi-modality, which is possibly due to a mixture of relatively smaller subgraphs

of real world graphs using our HADI algorithm, which will be described at the next section, and show the results in Figs. 3.1 ∼ 3.5. We have the following observations.

Diameter. What is the diameter of the Web? Albert et al. [2] computed the diameter on a directed Web graph with ≈0.3 million nodes, and conjectured that it is around 19 for the 1.4 billion-node Web graph. Broder et al. [6] used sampling from a ≈ 200 million-nodes Web graph and reported 16.15 and 6.83 as the diameter for the directed and the undirected cases, respectively. What should be the effective diameter, for a significantly larger crawl of the Web, with billions of nodes? Figure 3.1 gives the surprising answer:

Observation 1 (Small Web). *The effective diameter of the YahooWeb graph (year: 2002) is surprisingly small (≈ 7 ∼ 8).*

The previous results from Albert et al. and Broder et al. are based on the average diameter. For this reason, we also computed the average diameter and show the comparison of diameters of different graphs in Fig. 3.2. We first observe that the average diameters of all graphs are relatively small (<20) for both the directed and the undirected cases. We also observe that the Albert et al.'s conjecture of the diameter of the directed graph is over-pessimistic: both the sampling and HADI reported smaller values for the diameter of the directed graph. For the diameter of the undirected graph, we observe the constant or shrinking diameter pattern [27].

Static Radius Plot. How are the radii distributed in real networks? Is it Poisson? Lognormal? Figure 3.1 gives the surprising answer: multimodal! In other relatively

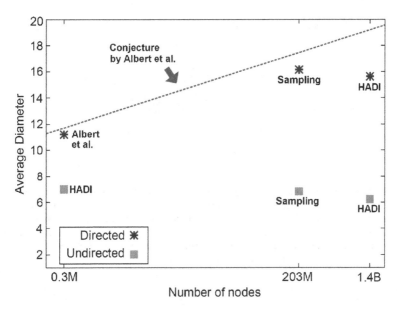

Fig. 3.2 Average diameter vs. number of nodes in lin-log scale for the three different Web graphs, where M and B represent millions and billions, respectively. (0.3M): Web pages inside nd.edu at 1999, from Albert et al.'s work [2]. (203M): Web pages crawled by Altavista at 1999, from Broder et al.'s work [6]. (1.4B): Web pages crawled by Yahoo at 2002 (YahooWeb in Table 3.1). The annotations (Albert et al., Sampling, HADI) near the points represent the algorithms for computing the diameter. The Albert et al.'s algorithm seems to be an exact breadth first search, although not clearly specified in their paper. Notice the relatively small diameters for both the directed and the undirected cases. Also notice that the diameters of the undirected Web graphs remain near-constant

small networks, however, we observe bi-modal structures. As shown in the Radius plot of U.S. Patent in Fig. 3.3a, they have a peak at zero, a dip at a small radius value (9) and another peak very close to the dip. Given the prevalence of the bi-modal shape, our conjecture is that the multi-modal shape of YahooWeb is possibly due to a mixture of relatively smaller sub-graphs, which got loosely connected recently.

Observation 2 (Multi-modal and Bi-modal). *The Radius distribution of the Web graph has a multi-modal structure. Many smaller networks have the bi-modal structure.*

About the bi-modal structure, a natural question to ask is what are the common properties of the nodes that belong to the first peak; similarly, for the nodes in the first dip, and the same for the nodes of the second peak. After investigation, the former are nodes that belong to the disconnected components (DCs); nodes in the dip are usually core nodes in the giant connected component (GCC), and the nodes at the second peak are the vast majority of well connected nodes in the GCC. Figure 3.3b exactly shows the radii distribution for the nodes of the GCC and the nodes of the few largest remaining components.

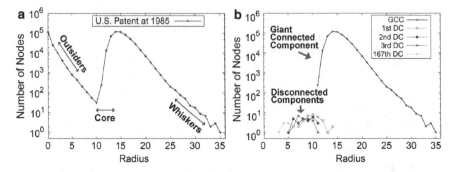

Fig. 3.3 (**a**) Static radius plot (count versus radius) of U.S. Patent graph. Notice the bi-modal structure with 'outsiders' (nodes in the disconnected components), 'core' (central nodes in the giant connected component), and 'whiskers' (nodes connected to the giant connected component with long paths). (**b**) The decomposition of the radius plot using the connected components information. Biggest curve with radius ranging from 11 to 35 the distribution for the giant connected component; small curves on the bottom, left several disconnected components

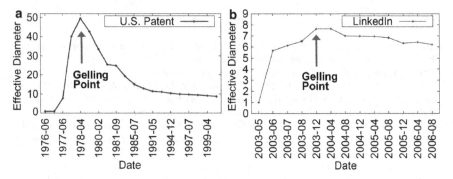

Fig. 3.4 Evolution of the effective diameter of real graphs. The diameter increases until a 'gelling' point, and starts to shrink after the point. (**a**) Patent. (**b**) LinkedIn

Now we can explain the three important areas of Fig. 3.3a: '*outsiders*' are the nodes in the disconnected components, and responsible for the first peak and the negative slope to the dip. '*Core*' are the central nodes with the smallest radii from the giant connected component. '*Whiskers*' [28] are the nodes connected to the GCC with long paths, and are the reasons of the second negative slope.

Dynamic Radius Plot. We study how the radius distribution changes over time. We know that the diameter of a graph typically grows with time, spikes at the 'gelling point', and then shrinks [27,31]. Indeed, this holds for our dataset, as shown in Fig. 3.4.

The question is, how does the radius distribution change over time? Does it still have the bi-modal pattern? Do the peaks and slopes change over time? We show the answer in Fig. 3.5. Notice that while the radius plot maintains its bi-modal shape,

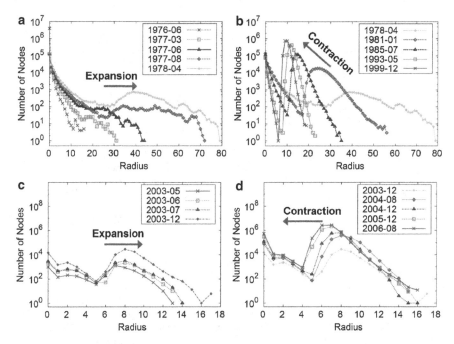

Fig. 3.5 Radius distribution over time. "Expansion": the radius distribution moves to the right until the gelling point. "Contraction": the radius distribution moves to the left after the gelling point. (**a**) Patent-expansion. (**b**) Patent-contraction. (**c**) LinkedIn-expansion. (**d**) LinkedIn-contraction

its width changes over time: it expands to the right until the 'gelling point', and it contracts after the gelling point. This clearly shows the correlation between the evolution of the diameter and the radius plot.

Observation 3 (Expansion-Contraction). *The radius distribution expands to the right until it reaches the gelling point. Then, it contracts to the left.*

3.3.2 Connected Components

A connected component of an undirected graph is a subgraph in which any two vertices are connected to each other by paths, and which is connected to no additional vertices. *What are the patterns and anomalies in the connected components of real world graphs?* With PEGASUS, we analyze the connected components of real world graphs, including the LinkedIn social network, Wikipedia page-linking-to-page network and the YahooWeb graph. Figure 3.6 shows the evolution of connected components of LinkedIn and Wikipedia graphs. Figure 3.7

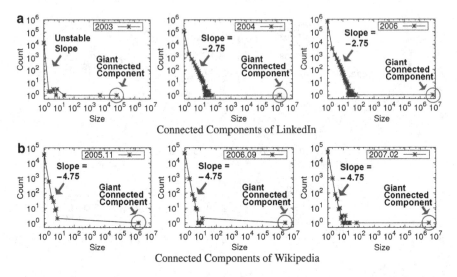

Fig. 3.6 The evolution of connected components. (**a**) The giant connected component grows for each year. However, the second largest connected component do not grow above Dunbar's number (≈ 150) and the slope of the size distribution remains constant after the gelling point at year 2003. (**b**) As in LinkedIn, notice the growth of giant connected component, the size of the second largest connected component bounded above, and the constant slope of the size distribution

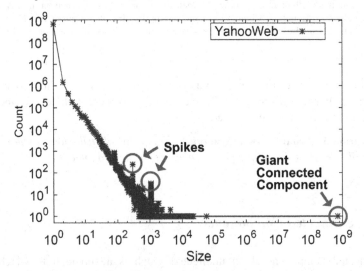

Fig. 3.7 Connected components size distribution of YahooWeb. Notice the two anomalous spikes which deviate significantly from its neighbors

shows the distribution of connected components in the YahooWeb graph. We have the following observations.

Power Laws in Connected Components Distributions. We observe the power law relation of count and size of small connected components in Figs. 3.6 and 3.7. This reflects that the connected components in real networks are formed by processes similar to Chinese Restaurant Process and Yule distribution [32].

Stable Connected Components After Gelling Point. In Fig. 3.6a, the distributions of connected components remain stable after a 'gelling' point [31] at 2003: the slopes of the size distributions do not change after the point. We observe the same phenomenon in Wikipedia graph in Fig. 3.6b. The graph shows stable slopes from the beginning, since the network were already mature in year 2005.

Absorbed Connected Components and Dunbar's Number. In Fig. 3.6a, we find two large connected components in year 2003. However it became merged in year 2004. The giant connected component keeps growing over time, while the second and the third largest connected components do not grow beyond size 100 until they are absorbed to the giant connected component. This agrees with the observation [31] that the sizes of the second/third connected components remain constant or oscillate, and the Dunbar's number [11], which says that the maximum community size in social networks is roughly 150. The connected components of Wikipedia in Fig. 3.6b also show that the sizes of the second/third connected components remain constant or oscillate.

Anomalous Connected Components. Figure 3.7 shows two outstanding spikes which deviate from the 'power-law' like size distributions of small disconnected components. In the first spike at size 300, more than half of the components have exactly the same structure and they were made from a domain selling company where each component represents a domain to be sold. The spike happened because the company *replicated* sites using the same template, and injected the disconnected components into the Web. In the second spike at size 1101, more than 80% of the components are adult sites disconnected from the giant connected component. Again, the adult sites are generated from a template. In sum, the distribution plot of connected components reveals interesting communities with special purposes, which are disconnected from the rest of the Internet.

3.3.3 Triangle Counting

Triangle in a graph is defined to be three nodes which are connected to each other. *What are the patterns and anomalies in the triangle counts, and the degrees in social network graphs?* Figure 3.8 shows the degree and the number of participating triangles in the Twitter 'who follows whom' graph at year 2009 [18]. We have the following observation which can be used to spot and eliminate harmful accounts, such as those of adult advertisers and spammers.

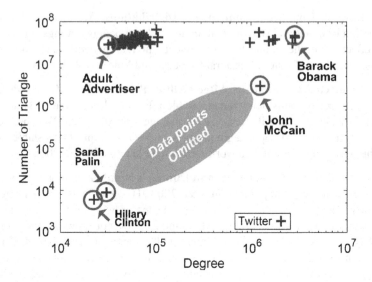

Fig. 3.8 The degree vs. participating triangles of some 'celebrities' in Twitter accounts. Also shown are accounts of adult sites advertisers which have smaller degree, but belong to an abnormally large number of triangles. The reason of the large number of triangles is that adult accounts are often created from the same provider, and they follow each other to form a clique, to possibly boost their rankings or popularity

Anomalous Triangles vs. Degree Ratio. In Fig. 3.8, celebrities have high degree and mildly connected followers, while adult sites advertisers have many fewer, but extremely well connected, followers, thereby creating a lot of triangles. The reason of the large number of triangles is that adult accounts are often created from the same provider, and they follow each other to possibly boost their rankings or popularity.

3.4 Algorithms

In this section, we describe the algorithms in the PEGASUS package, which enabled the discoveries in the previous section. First, we introduce GIM-V (Generalized Iterative Matrix-Vector multiplication), a general primitive including the diameter estimation and the connected component computation as special cases. Next, we describe the algorithm for large scale eigensolver.

3.4.1 Generalized Iterative Matrix Vector Multiplication

How can we find connected components, diameter, PageRank and node proximities of very large graphs quickly? How can we design a general primitive, which can be applied to many different algorithms? We show that, even though they seem unrelated, eventually we can unify them using the GIM-V primitive, standing for Generalized Iterative Matrix-Vector multiplication, which we describe in the next.

3.4.1.1 Main Idea

GIM-V is a generalization of normal matrix-vector multiplication. Suppose we have a n by n matrix M and a vector v of size n. Let $m_{i,j}$ denote the (i, j)-th element of M. Then the usual matrix-vector multiplication is

$$M \times v = v', \text{ where } v'_i = \sum_{j=1}^{n} m_{i,j} v_j$$

There are three operations in the previous formula, which, if generalized appropriately, provide a surprising number of useful graph mining algorithms:

1. combine2: multiply $m_{i,j}$ and v_j.
2. combineAll: sum n multiplication results for node i.
3. assign: overwrite the previous value of v_i with the new result to make v'_i.

GIM-V generalize the definition of the matrix-vector multiplication operator $\times$ to $\times_G$ where the three operations can be defined arbitrarily. Formally, we have:

$v' = M \times_G v$, where

$v'_i = $ assign $(v_i,$ combineAll$_i$ $(\{x_j \mid j{=}1..n,$ and $x_j{=}$ combine2 $(m_{i,j}, v_j)\}))$

The functions combine2(), combineAll() and assign() have the following signatures (generalizing the product, sum and assignment, respectively, that the traditional matrix-vector multiplication requires):

1. combine2$(m_{i,j}, v_j)$: combine $m_{i,j}$ and v_j.
2. combineAll$_i(x_1, \ldots, x_n)$: combine all results from combine2() for node i.
3. assign(v_i, v_{new}) : decide how to update v_i with v_{new}.

The 'Iterative' in the name of GIM-V denotes that the $\times_G$ operation is applied until an algorithm-specific convergence criterion is met. As we will see in a moment, by redefining these operations, we obtain different, useful algorithms including PageRank, Random Walk with Restart, connected components and diameter estimation. But first we want to highlight the strong connection of GIM-V with SQL. When combineAll$_i$() and assign() can be implemented by user defined functions, the operator $\times_G$ can be expressed concisely in terms of SQL. This viewpoint is important when we implement GIM-V in large scale parallel processing platforms, including Hadoop, if they can be customized to support several SQL primitives including JOIN and GROUP BY. Suppose that we have an edge table E(sid, did, val) and a vector table V(id, val), corresponding to a matrix and a vector, respectively. Then, $\times_G$ corresponds to the SQL statement in Table 3.2. We assume that we have (built-in or user-defined) functions, combineAll$_i$() and

Table 3.2 GIM-V in terms of SQL

SELECT E.sid, combineAll$_{E.sid}$(combine2(E.val,V.val)) FROM E, V WHERE E.did=V.id GROUP BY E.sid

combine2(), and we also assume that the resulting table/vector will be fed into the assign() function (omitted, for clarity).

In the following, we show how to customize GIM-V, to handle important graph mining operations including PageRank, Random Walk with Restart, diameter estimation and connected components.

3.4.1.2 GIM-V and PageRank

Our first application of GIM-V is PageRank, a famous algorithm used by Google to calculate relative importance of the Web pages [5]. The PageRank vector p of n Web pages satisfies the following eigenvector equation:

$$p = (cE^T + (1-c)U)p,$$

where c is a damping factor (usually set to 0.85), E is the row-normalized adjacency matrix, and U is a matrix with all elements set to $1/n$.

To calculate the eigenvector p we can use the power method, which multiplies an initial vector with the matrix, several times. We initialize the current PageRank vector p^{cur} and set all its elements to $1/n$. Then the next PageRank p^{next} is calculated by $p^{next} = (cE^T + (1-c)U)p^{cur}$. We continue to do the multiplication until p converges.

PageRank is a direct application of GIM-V. In this view, we first construct a matrix M by column-normalize E^T so that every column of M sums to 1. Then the next PageRank is calculated by $p^{next} = M \times_G p^{cur}$, where the three operations are defined as follows:

1. combine2$(m_{i,j}, v_j) = c \times m_{i,j} \times v_j$
2. combineAll$_i(x_1, \ldots, x_n) = \frac{(1-c)}{n} + \sum_{j=1}^{n} x_j$
3. assign$(v_i, v_{new}) = v_{new}$

3.4.1.3 GIM-V and Random Walk with Restart

Random Walk with Restart (RWR) is an algorithm to measure the proximity of nodes in a graph [35]. In RWR, the proximity vector r_k from node k satisfies the equation:

$$r_k = cMr_k + (1-c)e_k,$$

where e_k is a n-vector whose kth element is 1, and every other elements are 0. c is a restart probability parameter which is typically set to 0.85 [35]. M is the column-normalized and transposed adjacency matrix, as in Sect. 3.4.1.2. In GIM-V, RWR is formulated by $r_k^{next} = M \times_G r_k^{cur}$ where the three operations are defined as follows (δ_{ik} is the *Kronecker delta* which is 1 if $i = k$, and 0 otherwise):

1. $\text{combine2}(m_{i,j}, v_j) = c \times m_{i,j} \times v_j$
2. $\text{combineAll}_i(x_1, \ldots, x_n) = (1 - c)\delta_{ik} + \sum_{j=1}^{n} x_j$
3. $\text{assign}(v_i, v_{new}) = v_{new}$

3.4.1.4 GIM-V and Diameter/Radius Estimation

HADI[17, 19] is an algorithm to estimate the diameter and radius of large graphs. As described in Sect. 3.3.1, the diameter of a graph is the maximum of the length of the shortest paths between every pair of nodes, and the radius of a node v_i is the number of hops needed to reach the farthest-away node from v_i. The main idea of HADI is as follows. For each node v_i in the graph, we maintain the number of neighbors reachable from v_i within h hops. As h increases, the number of neighbors increases, until h reaches its maximum value. The diameter is h, when the number of neighbors within $h + 1$ hops does not increase for every node. For further details and optimizations, see [17, 19].

The main operation of HADI is updating the number of neighbors as h increases. Specifically, the number of neighbors of node v_i reachable within h hops is encoded in a probabilistic bitstring b_i^h which is updated as follows:

$$b_i^{h+1} = b_i^h \text{ BITWISE-OR } \{b_k^h \mid (i, k) \in E\}$$

In GIM-V, the bitstring update of HADI is expressed by

$$b^{h+1} = M \times_G b^h,$$

where M is the adjacency matrix, b^{h+1} is a vector of length n which is updated by $b_i^{h+1} = \text{assign}(b_i^h, \text{combineAll}_i(\{x_j \mid j = 1..n, \text{ and } x_j = \text{combine2}(m_{i,j}, b_j^h)\}))$, and the three operations are defined as follows:

1. $\text{combine2}(m_{i,j}, v_j) = m_{i,j} \times v_j$.
2. $\text{combineAll}_i(x_1, \ldots, x_n) = \text{BITWISE-OR}\{x_j \mid j = 1..n\}$
3. $\text{assign}(v_i, v_{new}) = \text{BITWISE-OR}(v_i, v_{new})$.

The $\times_G$ operation is run iteratively until the bitstrings for all the nodes converge.

3.4.1.5 GIM-V and Connected Components

GIM-V can compute connected components in large graphs. The main idea is as follows. For every node v_i in the graph, we maintain a component id c_i^h which is the minimum node id reachable from v_i within h hops. Initially, c_i^h of v_i is set to its own node id: that is, $c_i^0 = i$. For each iteration, each node sends its current c_i^h to its neighbors. Then c_i^{h+1}, component id of v_i at the next step, is set to the minimum value among its current component id and the received component ids from its neighbors. The crucial observation is that this communication between neighbors can be formulated in GIM-V as follows:

$$c^{h+1} = M \times_G c^h,$$

where M is the adjacency matrix, c^{h+1} is a vector of length n which is updated by $c_i^{h+1} = \text{assign}(c_i^h, \text{combineAll}_i(\{x_j \mid j = 1..n,$ and $x_j = \text{combine2}(m_{i,j}, c_j^h)\}))$, and the three operations are defined as follows:

1. $\text{combine2}(m_{i,j}, v_j) = m_{i,j} \times v_j$.
2. $\text{combineAll}_i(x_1, \ldots, x_n) = \text{MIN}\{x_j \mid j = 1..n\}$.
3. $\text{assign}(v_i, v_{new}) = \text{MIN}(v_i, v_{new})$.

By repeating this process, component ids of nodes in a component are set to the minimum node id of the component. We iteratively do the multiplication, until component ids converge. The upper bound of the number of iterations are determined by the following theorem.

Theorem 3.1 (Upper bound of iterations in GIM-V for Connected Components). *GIM-V for connected components requires maximum d iterations, where d is the diameter of the graph.*

Proof. The minimum node id is propagated to its neighbors at most d times.

Since the diameter of real graphs are relatively small, GIM-V for connected components completes after small number of iterations.

3.4.1.6   Fast Algorithms for GIM-V

Having defined GIM-V, the next challenge is to design efficient algorithms for computing it in MapReduce. We have the following three main ideas. First, we put together several nonzero elements into square blocks, and perform the block-wise matrix-vector multiplication instead of element-wise multiplication. Second, we cluster the graph so that nonzero elements in the adjacency matrix are closely located. Finally, we compress the nonzero bit strings of each block by standard compression algorithms like Gzip or Elias-γ [13]. This compression greatly saves space, which leads to faster running time of block-wise matrix-vector multiplication.

In the following, we describe our main ideas in detail by first describing the naive algorithm, and improving it by applying our main ideas one at a time.

GIM-V RAW: Naive Algorithm. In the naive algorithm for GIM-V, the inputs are an edge file and a vector file. Each line of the edge file contains one $(id_{src}, id_{dst}, mval)$, which corresponds to a non-zero cell in the adjacency matrix

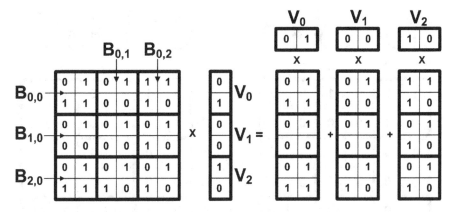

Fig. 3.9   GIM-V NNB. The matrix elements are grouped into 2×2 blocks denoted by $B_{i,j}$. The vector elements are grouped into length 2 blocks denoted by V_i. The matrix and vector are joined block-wise, not element-wise

M. Similarly, each line of the vector file contains one $(id, vval)$, which corresponds to an element in the vector V. The matrix-vector multiplication using these files is essentially a database-join: we group the elements that we want go together. For more details, please see [16, 20].

GIM-V NNB: Block Multiplication. GIM-V NNB adds block multiplication functionality to GIM-V RAW. The main idea is to group elements of the input matrix into blocks or submatrices of size b by b. Also we group elements of input vectors into blocks of length b. Here the grouping means we put all the elements in a group into one line of the input file. Each block contains only non-zero elements of the matrix or the vector. The format of a matrix block with k nonzero elements is $(row_{block}, col_{block}, row_{elem_1}, col_{elem_1}, mval_{elem_1}, \ldots, row_{elem_k}, col_{elem_k}, mval_{elem_k})$. Similarly, the format of a vector block with k nonzero elements is $(id_{block}, id_{elem_1}, vval_{elem_1}, \ldots, id_{elem_k}, vval_{elem_k})$. Only blocks with at least one nonzero elements are stored in disk. This block encoding forces nearby edges in the adjacency matrix to be closely located in disc; it is different from Hadoop's default behavior which do not guarantee co-locating them. After grouping, GIM-V is performed on blocks, not on individual elements. GIM-V NNB is illustrated in Fig. 3.9. This block encoding decreases the data size, and it leads to faster running time as shown in Fig. 3.13.

GIM-V NCB: Compression. GIM-V NCB further decreases the storage and the running time from GIM-V NNB by compressing the nonzero elements inside each block by standard compression algorithms like Gzip or Elias-γ.

GIM-V CCB: Clustering. GIM-V CCB even further increases performance of GIM-V NCB by clustering the nonzero elements of the adjacency matrix before the compression. In Fig. 3.10, the left and the right matrices come from two isomorphic graphs. However, the right matrix contains smaller number of denser blocks than

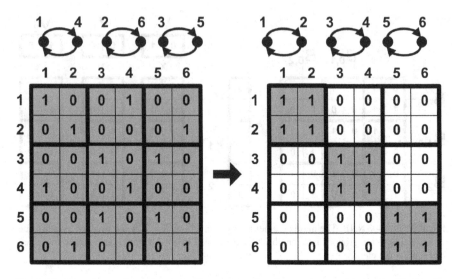

Fig. 3.10 Non-clustered vs. clustered adjacency matrices for two isomorphic graphs. Each node has a self loop which is omitted in the figure for clarity. The edges are grouped into 2 by 2 blocks. The right matrix uses only three blocks while the left matrix uses nine blocks. GIM-V CCB uses the clustered matrix

the left matrix, therefore can benefit more from compression algorithms. Clustering for large graphs form an active research area, and there are several existing works including Disco [36], Shingle ordering [7], and SlashBurn [15].

3.4.1.7   Performance

In this section, we provide the performance results, which show the effectiveness of our fast algorithms for GIM-V.

Machine Scalability. Figure 3.11 shows the scalability of GIM-V CCB method with regard to the number of machines. The Y-axis shows the 'scale up', that is, the ratio of the running time T_M with M machines, and T_{25}. We see that for all the graphs, the running time scales up near-linearly with the number of machines.

Edge Scalability. Figure 3.12 shows the scalability of CCB method with regard to the number of edges. We used the synthetic Kronecker graphs [26] for the experiments since graphs with any size can be easily generated. Note that for all the settings (10, 25, and 40 machines), the running time scales up near-linearly with the number of edges.

Effect of Optimization. Figure 3.13 shows the disk space and the running time comparisons of GIM-V variants. Note that the 'Proposed' CCB method, which combines the clustering and compression, provides up to 43× smaller storage and 9.2× faster running time compared to the naive NNB method.

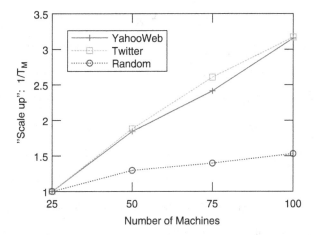

Fig. 3.11 Machine scalability of our proposed CCB method. The Y-axis shows the ratio of the running time T_M with M machines, and T_{25}, for PageRank queries. Note the running time scales up near-linearly with the number of machines

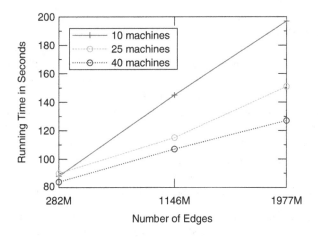

Fig. 3.12 Edge scalability of our proposed CCB method. The Y-axis shows the running time in seconds, for PageRank queries on Kronecker graphs. Note the running time scales up near-linearly with the number of edges for all the settings (10, 25, and 40 machines)

3.4.2   Eigensolver

Given a large graph, how can we find near-cliques (a set of tightly connected nodes) [38], the count of triangles [42], and related graph properties? They can be computed quickly provided we have the first several eigenvalues and eigenvectors of the adjacency matrix of the graph. In general, spectral analysis is a fundamental tool not only for graph mining, but also for other areas of data mining. Eigenvalues

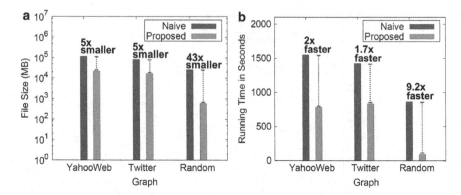

Fig. 3.13 Effectiveness of our proposed CCB method compared to the naive NNB method. (**a**) File size comparison after clustering and compression. The Y-axis is in log scale. Note our proposed method reduces the data size up to 43× smaller than the naive method. The 'Random' graph has better performance gain than real-world graphs since the density is much higher. (**b**) Running time comparison of PageRank queries. Our proposed method outperforms the naive method by 9.2×

and eigenvectors are at the heart of numerous algorithms, such as singular value decomposition (SVD) [4, 14], spectral clustering [33, 39], Principal Component Analysis (PCA) [37], Multi Dimensional Scaling (MDS) [3, 23], Latent Semantic Indexing (LSI) [9], and tensor analysis [12, 21, 22, 40]. Despite their importance, existing eigensolvers do not scale well: the maximum order and size of input matrices feasible for these solvers are million-scales.

To address this problem, we describe HEIGEN [18], a scalable algorithm for computing the top k eigenvalues and eigenvectors of matrices in MapReduce. HEIGEN can handle graphs with billions of nodes and edges easily. In designing HEIGEN, the major challenges are to carefully design algorithms that work well on distributed systems, and to exploit the inherent structure of data (e.g., skewness) in order to be efficient. We summarize the main ideas here and describe each in detail.

1. *Careful Algorithm Choice.* We carefully choose a sequential eigensolver algorithm that is efficient for MapReduce and gives accurate results.
2. *Selective Parallelization.* We group operations into expensive and inexpensive ones based on input sizes, and selectively parallelize them.
3. *Blocking.* We reduce the running time by decreasing the input data size and the amount of network traffic among machines using block encoding.
4. *Exploiting Skewness.* We decrease the running time by exploiting the skewness of data.

3.4.2.1 Careful Algorithm Choice

There are several sequential eigensolver algorithms. Which one should we choose for developing a scalable parallel eigensolver to compute top k eigenvalues and eigenvectors? We list the alternatives for computing the eigenvalues of a matrix and the reasoning behind our choice.

- **Power method**: the simplest and the most famous method for computing the topmost eigenvalue. However, it can not find the top k eigenvalues.
- **Simultaneous iteration (or QR)**: an extension of the Power method to find the top k eigenvalues. It requires large matrix-matrix multiplications that are prohibitively expensive for billion-scale graphs.
- **Lanczos-NO (No Orthogonalization)**: the basic Lanczos algorithm [25] which approximates the top k eigenvalues in the subspace composed of intermediate vectors from the Power method. The problem is that the computed eigenvalues can 'jump' up to larger eigenvalues: thus, the outputted eigenvalues might not be correct.
- **Lanczos-SO (Selective Orthogonalization)**: the problem of Lanczos-NO can be resolved by reorthogonalizing the current vector against all the previous intermediate vectors. However, this method requires many reorthogonalization. Lanczos-SO addresses this issue by selectively re-orthogonalizing vectors only when required, which we describe below in detail.

The Lanczos-SO has all the properties we need: it finds the top k largest eigenvalues and eigenvectors, it produces no spurious eigenvalues, and its most expensive operation, a matrix-vector multiplication, is tractable in MapReduce. Therefore, we select Lanczos-SO as our choice of the sequential algorithm for parallelization.

The main idea of Lanczos-SO is as follows: we start with a random initial basis vector b, which comprises a rank-1 subspace. For each iteration, a new basis vector is computed by multiplying the input matrix with the previous basis vector. The new basis vector is then orthogonalized against the last two basis vectors and is added to the previous rank-$(l-1)$ subspace, forming a rank-l subspace, where l is the number of the current iteration. Let Q_l be the $n \times l$ matrix whose i-th column is the i-th basis vector, and A be the matrix whose eigenvalues we seek to compute. We also define a $l \times l$ matrix $T_l = Q_l^* A Q_l$. Then, the eigenvalues of T_m are good approximations of the eigenvalues of A. Furthermore, multiplying Q_l by the eigenvectors of T_l provides a good approximation of the eigenvectors of A. We refer the interested reader to [41] for further details.

If we used the exact arithmetic, the newly computed basis vector would be orthogonal to all previous basis vectors. However, rounding errors from floating-point calculations compound and result in the loss of orthogonality. This is the cause of the spurious eigenvalues in Lanczos-NO. Orthogonality can be recovered once the new basis vector is fully re-orthogonalized to all previous vectors. However, this operation is quite expensive as it requires $O(l^2)$ re-orthogonalizations, where l is the

---

Algorithm 1: Lanczos-SO (selective orthogonalization)

Input: Matrix $A^{n \times n}$, random n-vector b, maximum number of steps l, error threshold ϵ, number of eigenvalues k

Output: Top k eigenvalues $\lambda_{1..k}$, eigenvectors $U^{n \times k}$

1: $\beta_0 \leftarrow 0$, $v_0 \leftarrow 0$, $v_1 \leftarrow b/||b||$;

2: **for** $i = 1..l$ **do**

3:      $v \leftarrow Av_i$; // Find a new basis vector

4:      $\alpha_i \leftarrow v_i^T v$;

5:      $v \leftarrow v - \beta_{i-1}v_{i-1} - \alpha_i v_i$; // Orthogonalize against two previous basis vectors

6:      $\beta_i \leftarrow ||v||$;

7:      $T_i \leftarrow$ (build tri-diagonal matrix from α and β);

8:      $QDQ^T \leftarrow EIG(T_i)$; // Eigen decomposition of T_i

9:      **for** $j = 1..i$ **do**

10:         **if** $\beta_i|Q[i, j]| \leq \sqrt{\epsilon}||T_i||$ **then**

11:             $r \leftarrow V_i Q[:, j]$;

12:             $v \leftarrow v - (r^T v)r$; // Selectively orthogonalize

13:      **if** (v was selectively orthogonalized) **then**

14:         $\beta_i \leftarrow ||v||$; // Recompute normalization constant β_i

15:      **if** $\beta_i = 0$ **then**

16:         break for loop;

17:      $v_{i+1} \leftarrow v/\beta_i$;

18: $T \leftarrow$ (build tri-diagonal matrix from α and β);

19: $QDQ^T \leftarrow EIG(T)$; // Eigen decomposition of T

20: $\lambda_{1..k} \leftarrow$ top k diagonal elements of D; // Compute eigenvalues

21: $U \leftarrow V_l Q_k$; // Compute eigenvectors. Q_k is the set of columns of Q corresponding to $\lambda_{1..k}$

---

number of iterations. A faster approach uses a quick test (line 10 of Algorithm 1) to selectively choose vectors that need to be re-orthogonalized to the new basis [10]. This selective-reorthogonalization idea is shown in Algorithm 1.

3.4.2.2 Selective Parallelization

Among many sub-operations in Algorithm 1, which operations should we parallelize? A naive approach is to parallelize all the operations; however, some operations run more quickly on a single machine rather than on multiple machines in parallel. The reason is that the overhead incurred by using MapReduce exceeds gains made by parallelizing the task; simple tasks where the input data is very small are carried out faster on a single machine. Thus, we divide the sub-operations into two groups: those to be parallelized and those to be run in a single machine. Table 3.3 summarizes our choice for each sub-operation. Note that the last two operations in Table 3.3 can be done with a single-machine standard eigensolver since the input matrices are tiny; they have l rows and columns, where l is the number of iterations.

Table 3.3 Parallelization choices. The last column of the table indicates whether the operation is parallelized in HEIGEN. Some operations are better to be run in parallel, since the input size is very large, while others are better in a single machine, since the input size is small and the overhead of parallel execution overshadows its decreased running time

Operation	Description	Input	(P?)
$y \leftarrow y + ax$	Vector update	Large	Yes
$\gamma \leftarrow x^T x$	Vector dot product	Large	Yes
$y \leftarrow \alpha y$	Vector scale	Large	Yes
$\|y\|$	Vector L2 norm	Large	Yes
$y \leftarrow M^{n \times n} x$	Large matrix-large, dense vector multiplication	Large	Yes
$y \leftarrow M_s^{n \times l} x_s$	Large matrix-small vector multiplication ($n \gg l$)	Large	Yes
$A_s \leftarrow M_s^{n \times l} N_s^{l \times k}$	Large matrix-small matrix multiplication ($n \gg l > k$)	Large	Yes
$\|T\|$	Matrix L2 norm which is the largest singular value of the matrix	Tiny	No
$EIG(T)$	Symmetric eigen decomposition to output QDQ^T	Tiny	No

3.4.2.3 Blocking

Minimizing the volume of information exchanged between nodes is important to designing efficient distributed algorithms. In HEIGEN, we decrease the amount of network traffic by using the block-based operations which is introduced in Sect. 3.4.1.6. Normally, one would put each edge "(source, destination)" in one line; Hadoop treats each line as a data element for its mapper functions. Instead, we propose to divide the adjacency matrix into blocks (and, of course, the corresponding vectors also into blocks), and put the edges of each block on a single line, and compress the source- and destination-ids. This makes the mapper functions a bit more complicated to process blocks, but it saves significant transfer time of data over the network. We use these edge-blocks and the vector-blocks for many parallel operations in Table 3.3, including matrix-vector multiplication, vector update, vector dot product, vector scale, and vector L2 norm. Performing operations on blocks is faster than on individual elements, since both the input size and the key space decrease. This reduces the network traffic and sorting time in the MapReduce shuffle stage.

3.4.2.4 Exploiting Skewness: Matrix-Vector Multiplication

In this section, we describe how HEIGEN implements matrix-vector multiplication algorithms by exploiting the skewness pattern of the data. There are two matrix-vector multiplication operations in Algorithm 1: the one with a large vector (line 3) and the other with a small vector (line 11).

The first matrix-vector operation multiplies a large matrix with a large and dense vector, and thus it requires a two-stage standard MapReduce algorithm by Kang et al. [16, 20]. In the first stage, matrix elements and vector elements are joined and multiplied to produce partial results. The partial results are added together to get the final result vector in the second stage.

Algorithm 2: CBMV (cache-based matrix-vector multiplication) for HEIGEN

Input: Matrix $M = \{(id_{src}, (id_{dst}, mval))\}$, Vector $x = \{(id, vval)\}$
Output: Result vector y
1: Stage1-Map(key k, value v, Vector x) // Multiply matrix elements and the vector x
2:   $id_{src} \leftarrow k$;
3:   $(id_{dst}, mval) \leftarrow v$;
4:   Output($id_{src}, (mval \times x[id_{dst}])$)); // Multiply and output partial results
5:
6: Stage1-Reduce(key k, values $V[]$) // Sum up partial results
7:   $sum \leftarrow 0$;
8:   **for** $v \in V$ **do**
9:       $sum \leftarrow sum + v$;
10: Output(k, sum);

The other matrix-vector operation, however, multiplies a large matrix with a small vector. HEIGEN uses the fact that the small vector can fit in a machine's main memory, and distributes the small vector to all the mappers using the distributed cache functionality of Hadoop. The advantage of the small vector being available in mappers is that joining edge elements and vector elements can be done inside the mapper, and thus the first stage of the standard two-stage matrix-vector multiplication algorithm can be omitted. In this one-stage algorithm the mapper joins matrix elements and vector elements to make partial results, and the reducer adds up the partial results.

The pseudo code of this algorithm, which we call CBMV (Cache-Based Matrix-Vector multiplication), is shown in Algorithm 2. We want to emphasize that this operation cannot be performed when the vector is large, as is the case of the first matrix-vector multiplication (line 3 of Algorithm 1). The CBMV is much faster than the standard method, as we will see in Sect. 3.4.2.6.

3.4.2.5   Exploiting Skewness: Matrix-Matrix Multiplication

Skewness can also be exploited to efficiently perform matrix-matrix multiplication (line 26 of Algorithm 1). In general, matrix-matrix multiplication is very expensive. A standard, yet naive, way of multiplying two matrices A and B in MapReduce is to join $A[i, :]$ and $B[:, j]$ for all pairs of (i, j) using a two-stage MapReduce algorithm. This algorithm, which we call MM (direct Matrix-Matrix multiplication), is very inefficient since it generates a huge intermediate data in the shuffle stage of MapReduce. Fortunately, when one of the matrices is very small, we can exploit the skewness to come up with an efficient MapReduce algorithm. This is exactly the case in HEIGEN; the first matrix is very large, and the second is very small.

The main idea is to distribute the second matrix using the distributed cache functionality in Hadoop, and multiply each element of the first matrix with the corresponding row of the second matrix. We call the resulting algorithm Cache-Based Matrix-Matrix multiplication, or CBMM.

Fig. 3.14 Comparison of running time between different skewed matrix-matrix multiplication methods in MapReduce. Our proposed CBMM outperforms naive methods by at least 76×. The slowest matrix-matrix multiplication algorithm (MM) even didn't finish and the job failed due to the excessive amount of data

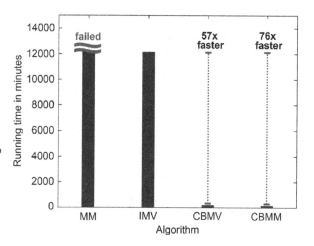

There are other alternatives to matrix-matrix multiplication: one can decompose the second matrix into column vectors and iteratively multiply the first matrix with each of these vectors. We call the algorithms, introduced in Sect. 3.4.2.4, as *Iterative matrix-vector multiplications* (IMV) and *Cache-based iterative matrix-vector multiplications* (CBMV). The difference between CBMV and IMV is that CBMV uses cache-based operations, while IMV does not. As we will see in Sect. 3.4.2.6, the best method (CBMM) is faster than naive methods.

3.4.2.6 Performance

We show the performance result showing the effectiveness of our fast algorithms for HEIGEN. Figure 3.14 shows the running time comparison of different skewed matrix-matrix multiplication methods in MapReduce. Note that cache-based methods (CBMM and CBMV) outperform other methods by at least 57×. The best method CBMM outperforms naive methods by at least 76×. The slowest matrix-matrix multiplication algorithm (MM) even didn't finish and the job failed because of the excessive amount of data.

3.5 Conclusion

In this chapter we presented PEGASUS, a graph mining library for finding patterns and anomalies in massive, real-world graphs. Our major contributions include:

- Scalable algorithms for mining billion-scale graphs.
- Performance analysis of our proposed method, which achieves up to 43× smaller storage and 9.2× faster running time.

- Discovery of patterns and anomalies of in huge, real-world graphs. Some of our most impressive findings are (a) the discovery of adult advertisers in the who-follows-whom on Twitter and (b) the 7-degrees of separation in the Web graph.

There are many real world graphs which have been kept intact. We expect to analyze them using PEGASUS, thereby transforming the massive raw data into valuable knowledge.

References

1. Aggarwal, G., Data, M., Rajagopalan, S., Ruhl, M.: On the streaming model augmented with a sorting primitive. In: Proceedings of FOCS, Rome (2004)
2. Albert, R., Jeong, H., Barabasi, A.L.: Diameter of the World Wide Web. Nature **401**, 130–131 (1999)
3. Bartell, B.T., Cottrell, G.W., Belew, R.K.: Latent semantic indexing is an optimal special case of multidimensional scaling. In: SIGIR, Copenhagen (1992)
4. Berry, M.W.: Large scale singular value computations. Int. J. Supercomput. Appl. **6**, 13–49 (1992)
5. Brin, S., Page, L.: The anatomy of a large-scale hypertextual (Web) search engine. In: WWW, Brisbane (1998)
6. Broder, A., Kumar, R., Maghoul, F., Raghavan, P., Rajagopalan, S., Stata, R., Tomkins, A., Wiener, J.: Graph structure in the Web. Comput. Netw. **33**, 309–320 (2000)
7. Chierichetti, F., Kumar, R., Lattanzi, S., Mitzenmacher, M., Panconesi, A., Raghavan, P.: On compressing social networks. In: KDD, Paris (2009)
8. Dean, J., Ghemawat, S.: MapReduce: simplified data processing on large clusters. In: OSDI, San Francisco (2004)
9. Deerwester, S., Dumais, S.T., Furnas, G.W., Landauer, T.K., Harshman, R.: Indexing by latent semantic analysis. J. Am. Soc. Inf. Sci. **41**(6), 391–407 (1990)
10. Demmel, J.W.: Applied Numerical Linear Algebra. SIAM, Philadelphia (1997)
11. Dunbar, R.: Grooming, Gossip, and the Evolution of Language. Harvard University Press, Cambridge (1998)
12. Dunlavy, D.M., Kolda, T.G., Acar, E.: Temporal link prediction using matrix and tensor factorizations. TKDD **5**(2), Article 10 (2011)
13. Elias, P.: Universal codeword sets and representations of the integers. IEEE Trans. Inf. Theory **21**(2), 194–203 (1975)
14. Kamel, M.: Computing the singular value decomposition in image processing. In: Proceedings of Conference on Information Systems, Tucson (1984)
15. Kang, U., Faloutsos, C.: Beyond 'caveman communities': hubs and spokes for graph compression and mining. In: ICDM, Vancouver (2011)
16. Kang, U., Tsourakakis, C.E., Faloutsos, C.: PEGASUS: a peta-scale graph mining system – implementation and observations. In: IEEE International Conference on Data Mining, Miami (2009)
17. Kang, U., Tsourakakis, C.E., Appel, A., Faloutsos, C., Leskovec, J.: Radius plots for mining tera-byte scale graphs: algorithms, patterns, and observations. In: SDM, Columbus (2010)
18. Kang, U., Meeder, B., Faloutsos, C.: Spectral analysis for billion-scale graphs: discoveries and implementation. In: PAKDD, Shenzhen (2011)
19. Kang, U., Tsourakakis, C.E., Appel, A., Faloutsos, C., Lekovec, J.: HADI: mining radii of large graphs. ACM Trans. Knowl. Disc. Data **5**, 8:1–8:24 (2011)
20. Kang, U., Tsourakakis, C.E., Faloutsos, C.: PEGASUS: mining peta-scale graphs. Knowl. Inf. Syst. **27**(2), 303–325 (2011)

21. Kolda, T.G., Bader, B.W.: Tensor decompositions and applications. SIAM Rev. **51**(3), 455–500 (2009)
22. Kolda, T.G., Sun, J.: Scalable tensor decompsitions for multi-aspect data mining. In: ICDM, Pisa (2008)
23. Kruskal, J.B., Wish, M.: Multidimensional Scaling. SAGE, Newbury Park (1978)
24. Lämmel, R.: Google's MapReduce programming model – revisited. Sci. Comput. Program. **70**, 1–30 (2008)
25. Lanczos, C.: An iteration method for the solution of the eigenvalue problem of linear differential and integral operators. J. Res. Nat. Bur. Stand. **45**, 255 (1950)
26. Leskovec, J., Chakrabarti, D., Kleinberg, J.M., Faloutsos, C.: Realistic, mathematically tractable graph generation and evolution, using kronecker multiplication. In: PKDD, Porto (2005)
27. Leskovec, J., Kleinberg, J., Faloutsos, C.: Graphs over time: densification laws, shrinking diameters and possible explanations. In: SIGKDD, Chicago (2005)
28. Leskovec, J., Lang, K.J., Dasgupta, A., Mahoney, M.W.: Statistical properties of community structure in large social and information networks. In: WWW, Beijing (2008)
29. Low, Y., Gonzalez, J., Kyrola, A., Bickson, D., Guestrin, C., Hellerstein, J.M.: GraphLab: a new framework for parallel machine learning. In: UAI, Catalina Island (2010)
30. Malewicz, G., Austern, M.H., Bik, A.J.C., Dehnert, J.C., Horn, I., Leiser, N., Czajkowski, G.: Pregel: a system for large-scale graph processing. In: SIGMOD Conference, Indianapolis (2010)
31. Mcglohon, M., Akoglu, L., Faloutsos, C.: Weighted graphs and disconnected components: patterns and a generator. In: KDD, Las Vegas (2008)
32. Newman, M.E.J.: Power laws, Pareto distributions and Zipf's law. Contemp. Phys. **46**, 323–351 (2005)
33. Ng, A.Y., Jordan, M.I., Weiss, Y.: On spectral clustering: analysis and an algorithm. In: IPS, Vancouver (2002)
34. Olston, C., Reed, B., Srivastava, U., Kumar, R., Tomkins, A.: Pig latin: a not-so-foreign language for data processing. In: SIGMOD, Vancouver (2008)
35. Pan, J., Yang, H., Faloutsos, C., Duygulu, P.: Automatic multimedia cross-modal correlation discovery. In: KDD, Seattle (2004)
36. Papadimitriou, S., Sun, J.: DisCo: distributed co-clustering with Map-Reduce. In: IEEE International Conference on Data Mining, Pisa (2008)
37. Pearson, K.: On lines and planes of closest fit to systems of points in space. Philos. Mag. **2**(6), 559–572 (1901)
38. Prakash, B.A., Seshadri, M., Sridharan, A., Machiraju, S., Faloutsos, C.: EigenSpokes: surprising patterns and community structure in large graphs. In: PAKDD, Hyderabad (2010)
39. Shi, J., Malik, J.: Normalized cuts and image segmentation. In: CVPR, San Juan (1997)
40. Sun, J., Tao, D., Faloutsos, C.: Beyond streams and graphs: dynamic tensor analysis. In: KDD, Philadelphia (2006)
41. Trefethen, L.N., Bau III, D.: Numerical Linear Algebra. SIAM, Philadelphia (1997)
42. Tsourakakis, C.E.: Fast counting of triangles in large real networks without counting: algorithms and laws. In: ICDM, Pisa (2008)

Chapter 4
Customer Analyst for the Telecom Industry

David Konopnicki and Michal Shmueli-Scheuer

Abstract The telecommunications industry is particularly rich in customer data, and telecom companies want to use this data to prevent customer churn, and improve the revenue per user through personalization and customer acquisition. Massive-scale analytics tools provide an opportunity to achieve this in is a flexible and scalable way. In this context, we have developed IBM Customer Analyst, a components library to analyze customer behavioral data and enable new insights and business scenarios based on the analysis of the relationship between users and the content they create and consume. Due to the massive amount of data and large number of users, this technology is built on IBM Infosphere BigInsights and Apache Hadoop. In this work, we first describe an efficient user profiling framework, with high user profiling quality guarantees, based on mobile web browsing log analysis. We describe the use of the Open Directory Project categories to generate user profiles. We then describe an end-to-end analysis flow and discuss its challenges. Last, we validate our methods through extensive experiments based on real data sets.

4.1 Introduction

The Telecommunications industry (telecom) seems to be one of the best fitted candidates to adopt *Massive-Scale Analytics* (MSA) technologies. In order to understand why it is so, we first must have a brief look at the history of what is now called MSA technologies and their development and how, in this historical perspective, the world of telecom was meant to be deeply affected.

D. Konopnicki (✉) • M. Shmueli-Scheuer
IBM Haifa Research Lab, Haifa, Israel
e-mail: davidko@il.ibm.com; shmueli@il.ibm.com

A. Gkoulalas-Divanis and A. Labbi (eds.), *Large-Scale Data Analytics*,
DOI 10.1007/978-1-4614-9242-9_4, © Springer Science+Business Media New York 2014

4.1.1   How It All Began

Clearly, the problem of managing and computing over very large datasets exists for a very long time. In some sense, the Online Transaction Processing Systems, Grid Computing systems or, even supercomputers, all being developed for decades, can be considered as precursors of the modern MSA systems. However, if we need to define a historical starting point to the MSA phenomena, the publication by Google's researcher's seminal articles on the Map-Reduce paradigm and the Google File System should certainly be it [7, 9]. These articles suggested that huge computing tasks, like those involved in the refresh of the Google Search Index, can be realized, in a reliable fashion, on a very large infrastructure of relatively cheap computers: the infrastructure was to liberate programmers from two underlying very complex issues, namely the complexity of writing parallel processing code and the inherent unreliability of hardware.

Even if the originality of the approach is still being discussed, one cannot ignore the impact this approach had. At a time when the computer industry was fascinated by Google's technical prowess, in addition to its financial results, it seemed suddenly that its technology was no magic and that, on the contrary, any company could quickly benefit from the same tools. In fact, it did not take a long time for the open-source community, aided by some of Google's competitors, to develop the Apache Hadoop software framework (Hadoop) and the Hadoop File System (HFS), mimicking Google's infrastructure.

Massive-scale analytic technologies have historically stemmed from the Internet dominating companies, like Google and Facebook. Providing free or cheap services to the masses, those companies' business model is usually based on the analysis of the behavior of their users in order to provide personalized advertisement, content recommendations and other personalized services. Thus, the first use-cases for the MSA technologies and, still, some of the more convincing, have been related to the analysis of customers' data and personalization of services, and continue to be so.

Beside the centrality of customer data analysis scenarios, an important technical feature of MSA should be noted for the implications it has on the technology applicability. As it has been built initially to deal with managing search indexes, MSA technology has the particularity of being schema-less or, at least, schema-light. The meaning of this property is that MSA technologies are particularly appropriate for dealing with data whose format changes quickly, for dealing with data integration from different sources and, for dealing with non-traditional data types, like text and graphs. This is in contrast to relational database technologies, for which the problems of schema definition, evolution and data integration, have always been acute.

It is important to notice that this technical particularity makes MSA technology particularly relevant to customer data analysis. The reason is that there are multiple interaction points between a company and its customers. For example, customers interact with a company IT systems through different stores, different customer management systems (online and offline) and more. Those IT systems are usually organized in silos and some involve structured data, like transactions lists, while others involve textual data (e.g., letters, conversations). Thus, as surprising as it may sound, it has always been a challenge for companies to monitor and manage a consistent and encompassing view of their customer base and all interaction channels. This is where the MSA technology comes into play; by allowing to integrate, in a flexible way, different data sets and enabling to cope with both structured and unstructured data, MSA technologies bring the promise of enabling the long expected 360° view of the customers.

4.1.2 The Telecom Role and Problem

MSA technologies have nothing that is specifically designed for telecom applications. Indeed, MSA is appropriate for all industries that can benefit from the analysis of the behavior of individual customers; the retail industry, the banking and finance industry, the media industry, are not less relevant. It is some other elements that make MSA fit with telecom particularly well.

The telecom industry has always been rich in customer information: the customer base of its biggest companies span the population of entire countries, while billing purposes require companies to monitor the activity of its customers. From the 1990s and on, the development of the cellular phones technology and its very large adoption, expanded the range of customer information available to telecom companies to new types of data (e.g., location information). Moreover, the development of the mobile internet access, paved the way for telecom companies to become also content providers and Internet Service Providers, hence, giving them access to even more customer data.

On the other hand, following years of surge and profits, the telecom companies have been more and more pressured by the markets. Due to fierce competition, the saturation of their customer market and legislation enforcement, they have seen their revenue and competitivity decrease. Paradoxically, the age of ubiquity of the cellular phones and their raising importance in almost all human interactions, is also the age of doubt and uncertainty for telecom companies.

One particular reason to the decline of telecom companies brings us back to the intersection of business and technology, and back to MSA. Paradoxically, while 70% and more of the earth population is in possession of a cellular phone and, hence, customers of telecom companies, the business model of these companies is undermined by the giant computer and internet companies, like Apple, Google, Amazon and Facebook. Through cellular phones, Apple, Google, Amazon and Facebook control more and more devices, operating systems and applications and, thus, user time. Hence, they are more and more in control of the customer interactions and opportunity for monetization of those interactions through advertisement, merchandising and more. Apple, Google, Amazon and Facebook do this by mastering the MSA technologies providing an unprecendented personalized experience to their customers.

As a result, telecom companies want and need to compete with those computing giants by excelling in the domain of customer data analysis, through the use of MSA technologies.

4.1.3 Possible Telecom Data Sets

After having explained why the MSA technologies are particularly important to the telecom world, we present an overview of the large datasets managed by telecom companies that can be used as a base for customer data analysis. As we mentioned earlier, telecom companies are particularly rich in data about their customers.

4.1.3.1 Identifying Users

Every Telecom company makes a fundamental distinction between two types of customers: pre-paid and post-paid customers. Post-paid customers are customers who are, typically, contractually linked to a particular Telecom company. They are able to consume voice calls and data with their devices and pay a phone bill at the end of each month. Pre-paid customers are customers who pay in advance for credits to a company of their choice and then can use those credits by use of phone services.

This fundamental economic distinction has a data aspect: since post-paid customers must be identified and located for the Telecom company to be confident that they will pay their bills, the registration of a post-paid Telecom customer involves the registration of the user identity, address and other, identifying or not, demographic attributes of the user in a database. Post-paid customers are usually linked to a particular Telecom company for several years. This allows the companies to gather huge behavioral data sets about those customers.

On the other hand, being a pre-paid customer does not require any kind of registration.[1] The user is free to buy credit from any Telecom company at any time. This type of customers is much more difficult to track. They move from company to company very frequently, depending on which one gives the best economical conditions at any given time.

4.1.3.2 Call/Events Data Records

The most fundamental pieces of information managed by Telecom companies are *Call Data Records* (CDRs). CDRs are emitted by telephony equipment in order to digitally record the details of a phone call passed through the device. CDRs

---

[1]Some countries in fact require some kind of registration to take place but it is not a technological or business requirement.

typically identify the caller and the receiver of a call, time and duration, together with various other parameters describing, for example, whether any fault condition has been encountered. It is important to note that, since the CDR data contain the identification of the base station from which a call has occurred, it is possible to process these records in order to localize cellular phone users.

During what one may consider as one "phone conversation", several such CDRs may be emitted by several different pieces of equipment and need to be retrieved and reconciled through a complex process, called *Mediation*. The reconciled CDRs, resulting from the mediation process, are subsequently used by billing systems to produce phone bills. In big telecom companies billions of CDRs are produced, treated and analyzed every day making CDRs the first source of data that necessitates MSA technologies.

Event Data/Detail Records (EDRs) are an extension of CDRs for systems that deal with more than voice calls. They are used to measure any type of events occurring in a telecom network: the sending of a message, web browsing activities, the download of a movie... In this chapter, we will present a detailed example of analyzing EDRs using MSA technologies.

4.1.3.3 The Apps World

One of the recent revolutions in the world of telecommunications is the emergence of the *Mobile Applications* (Apps) *mobile apps* as the main interaction channel between end-users and services. For several years, the Telecom world have struggled to find the right way to port the web experience to mobile devices. Several data formats have been proposed (e.g., HDML, WML, cHTML) together with technical solutions. However, usage has always been plagued by several limitations (small screen sizes, limited flexibility of navigation and more). The success of Apple's Iphone led to the emergence of the Apps, which are dedicated applications for each provider, as the preferred channel to content consumption (and more) on Mobile Devices (as opposed to the multiple purpose browser used for web browsing).

Telecom companies that wish to understand what end-users do with their phones, need to monitor apps usage. However, this is a daunting task: usually the apps are not distributed by the providers themselves and, while most Apps use a client-server architecture, their backend server do not make use of the telecom companies infrastructure. Thus, the only way to monitor apps usage is to study the communication between an app and its server when transmitted over the network.

This is a challenging task as every application uses a different communication protocol with its server. On the other hand, from a technical point of view, it still goes to analyzing EDRs.

4.1.3.4 Other Customer Data Sources

Other sources of data are available to telecom providers, some telecom specific, some general. Users interactions with the network are monitored by several systems: for example, the location of every end-user is monitored every few seconds so it is connected with the tower with the strongest signal.

Other sources of customer data include traditional Customer Relationship Management systems, web analytics tools (for monitoring the Telcom company's web site visitors) and more.

4.1.4 Some Applications

So far we have provided a high-level view of the data sources that are available for MSA analysis. In what follows, we discuss possible business uses of this data.

4.1.4.1 Churn Prevention

In a saturated mobile phone market, preventing customer churn is important to telecom companies. This need is exacerbated by the fact that, in lots of cases, telecom companies are actually sponsoring newly acquired users (e.g., by giving them free devices and rebates for calls), the meaning being that they begin to see a profit from user activities only after a few years! In addition, preventing the churn of an existing customer is much cheaper and has a higher return on investment than acquiring a new one.

Thus, churn prevention is a primary domain of focus for analytics in telecom companies: CDR data is analyzed in order to track potential indicators of churn. Those indicators might be related to a decrease in call frequency, a higher rate of network problem encountered by a particular users, or can even exploit the social network, as expressed by the CDR data [23]. In general, churn prevention models are built using Machine Learning techniques applied to the customer characteristics in order to understand which features can be used as early predictors of churn.

In this chapter, we present in detail how MSA analytics is applied to EDR data that can be analyzed, such that new users characteristics corresponding to personal interests can be computed. Clearly, these new characteristics can be included into churn models in order to discover customer churn induced by personal interest; for example, users churning to a competitor due to a sports fan targeted offering.

4.1.4.2 User Segmentation and Personalized Marketing

Marketing activities are often organized by companies according to some segmentation of their customer base. The meaning is that marketeers try to look at their customers as belonging to several groups having uniform characteristics. For example, students, aging urbans, etc. Then, by defining to which groups they want to market (e.g., prevent churn, sell new services), the segment being addressed helps defining the marketing message, the media used to reach customers and more. Many times, segments are built based on marketeers intuition or customer surveys.

One of the goals of customer data analysis is to find new segments of customers that should be targeted on marketing activities. Clearly, the analysis of EDR data that we present in this chapter can be used for such a task, as it allows to track users interests.

Furthermore, the customer experience on the web is getting more and more personalized as the Internet giants use MSA techniques to mine the personal data they have gathered about their customers (e.g., Amazon product recommendations). Hence, mining customer data can be realized by telecom companies in a similar way, in order to provide the same kind of experience and economical benefits.

Other applications to customer data analysis by telecom companies are possible: this includes the optimization of tiered pricing plans, personalized advertisement models and network optimizations. We cannot enter into all the details of those applications. It is sufficient to note that the more is known on customers and their behavior, the more it is possible to customize offers to them, hence potentially increasing acquisition, preventing churn and increasing the value of every customer.

4.1.5 Privacy Issues

Clearly, the analysis of customer data raises concerns about users' privacy. Such concerns are not the focus of this chapter. We refer the interested readers to [11] for an extensive review of social and technological aspects of privacy. It is also interesting to note the difference between the European view of privacy, as a domain of mainly government regulation and protection of personal data, as opposed to the more flexible American view of personal data, as owned by the individuals which are free to make commercial use of it. It is also useful to point the reader to the recent guidelines published by the *Electronic Frontier Foundation* on this subject [12].

4.1.6 Customer Analyst

In order to address the scenarios we described, we created *Customer Analyst*, a library of algorithms implemented as a set of components. In order to cope with massive-scale scenarios these components run on top of Apache Hadoop using the MapReduce paradigm. These components can be connected into analytics workflows in order to implement various scenarios.

In general, Customer Analyst deals with the relationships between end-users and content. Content can be text being consumed, e.g., web pages browsed, or it can content being produced, e.g., tweets or blog entries written by the end-user. The idea is that through the analysis of content associated with users, it is possible to model the users varying interests, including spatio-temporal correlations, and this can be used for micro-segmentation, marketing, advertising and more.

4.1.7 Chapter Organization

In the remainder of this chapter, we first describe an efficient user profiling framework with high quality guarantees. We describe the use of the ODP (Open Directory Project) categories to generate user profile. We then describe an end-to-end analysis flow and discuss its challenges. Last, we validate our methods through extensive experiments based on real data sets.

4.2 Related Work

In this chapter, we present our work on generating user profiles based on the ODP taxonomy for the telecom industry. Generating user profiles for the telecom industry has been intensively studied with respect to different scenarios. Customer churn, as discussed in Sect. 4.1.4.1, is one of the biggest challenges of the Telecom companies. In Richter et al. [23] the authors exploit the user social network based on the CDRs to detect users with high churn propensity. Hung et al. [13] predict the propensity to churn using decision trees and neural network models based on customer demographics, billing information, contract status, call detail records and service change logs.

Another important scenario is the so-called *location aware services*. Location aware services use the location of the mobile user to adapt services, including shopping, entertainment, traveling and more. Kaasinen [15] studied location-aware mobile services from the users' point of view, focusing on information needs while the user is on a move, delivery methods (push vs. pull) and more. De Reuver and Haaker [8] studied mobile business model concepts for the domain of context-aware services to increase the adoption of location-based services, where an important aspect is the management of user profiles that contains user interests, preferences and behavior. Van Setten et al. [27] suggested a context-aware mobile tourist application that adapts its services to the users' needs based on both the users' interests and their current context.

Some papers have utilized the ODP for different uses. The main focus has been on personalized search applications. Chirita et al. [4] focused on exploiting the ODP to achieve high quality personalized web search based on the distance of the categories of the returned URL to the user profile categories. The distance is measured by hierarchical semantics and the ODP tree structure. In [26] the authors apply the ODP to enhance the HITS algorithm [16] using dynamic user profiles.

Oishi et al. [20] created user profiles expressed as a category vector of the top most categories of the ODP. The authors then showed how to rerank web search results by using the ODP based the user profile. Other works exploit the ODP for labeling tasks, as described in [6, 19]. Davidov et al. [6] used the ODP in order to label datasets for text categorization by measuring conceptual distance between categories, where larger distance means easier dataset. In [19] the authors utilized the ODP for partitional hierarchical clustering algorithm. Specifically, the authors converted one branch of the ODP into relational database tables for efficient use in their clustering algorithm. Finally, there are some works about spam filtering using the ODP taxonomy [10, 28].

Recently, with the growth in the amount of online generated data and users, several works have focused on developing frameworks and algorithms for targeting users in large scale systems [1, 3, 24, 29]. Shmueli-Scheuer et al. [24] suggested a large scale user profiling framework using Hadoop MapReduce. Their focus was on generating profiles based on the content of web pages and representing the user profiles as the set of terms that most differentiate the user from the rest of the population. Chen et al. [3] developed a behavioral targeting system over the Hadoop MapReduce framework to select the ads that are most relevant to users. Zhou et al. [29] proposed large scale collaborative filtering techniques for movies recommendation. Finally, Cetintemel et al. [1] suggested an incremental algorithm for constructing user profiles based on monitoring and explicit user feedback. Their approach allows to trade between the profile complexity and its quality. In the context of these works, in this paper we describe how a very large number of content-based profiles can be extracted and maintained using the Apache Hadoop MapReduce framework.

4.3 Customer Modeling Framework

In what follows, we describe the details of a large-scale telecom user profiling framework. We begin by describing the general setting, followed by the two-phases of user profile modeling, namely, *categorization* and *aggregation*. We conclude this section with a description of the implementation on the modeling algorithm using the MapReduce paradigm.

4.3.1 Setting

In this section, we describe the inputs to our analysis flow. As described in the introduction, telecom companies monitor the data traffic that traverses their systems. In particular, they log each HTTP request in a system log, containing all users' interactions with web pages (documents). This log also contains Mobile Apps interactions, whose importance we have stressed earlier. However, for the sake of simplicity, we will focus now on the interactions with web pages, while it should be easy to the reader to understand how this work can be extended to Mobile Apps. An additional constraint that must be taken into account is that we do not have access to the full URLs of the documents but only the domain of the URLs due to privacy reasons.

Consequently, each record in the input log is a tuple $\langle u, d, context \rangle$ and captures a single user-document association, where u represents the user, d represents the domain-level URL (formally defined below Sect. 4.3.2.1) of the document associated with that user, and *context* is any additional available metadata extracted from the context of the user-document association (e.g., time, date, geographic location of the user, user agent, content type, etc.). A user profiling module consumes the system log data at scheduled time periods (e.g., once a day) and is responsible to maintain user profiles. Both the system log and user profiles are stored in Apache Hadoop's distributed file system (HDFS).

4.3.2   The User Profile Model

Telecom companies manage user profiles that are based on structured data, such as user demographics, analysis of voice CDRs, network data metrics and more. Customer Analyst aims at enriching those existing user profiles based on unstructured information by analyzing web browsing activities.

For example, a user that accesses nba.com very often, is likely to be a basketball fan. In order to generate the user profile based on web browsing, a systematic approach is needed. Specifically, an approach that (a) categorizes the different pages effectively transforming opaque URLs into meaningful data, and (b) aggregates them wisely into a comprehensive user profile.

As explained above, those user profiles can then be used by telecom companies for targeted services to users (e.g., advertisements, recommendations) or to generate new micro-segments based on their web behavior, improve churn prediction models and more. In the following sections, we will describe the two phases of generation of user profiles.

4.3.2.1   Basic Notations

In the World Wide Web, every web page is identified by a unique URL, which usually consists of a schema, domain name, port number and the path to be fetched, with the syntax scheme://domain:port/path. The scheme name defines the namespace, the domain name gives the destination location for the URL, the optional port number completes the destination address for a communications session, and finally, the path is used to specify the local resource requested. For simplicity, let us assume that the port number is always defaulted. Let us define *domain-level* URL to be a URL that is of the form scheme://domain, denoted as d, and *URL-level* URL to be scheme://domain/path, denoted as u. Equipped with this terminology we define the first phase, the categorization.

4.3.2.2 Categorization

There are different ways to model web pages data; common approaches (e.g., [5, 22]) work by extracting the content of the web pages, together with some other optional metadata, such as title, hyperlinks, layout, etc., and then categorize those pages based on this information. A different approach is to use pre-categorized data, e.g., the DMOZ Open Directory Project (ODP).[2] This is the approach that is used by our user profiling algorithms.

The ODP is one of the largest collaborative sources that contains manually annotated web pages. This effort resulted in the categorization of more than four million web pages into more than 590,000 categories (such as Arts, Business, Computers, Games, Health, News, etc.), expressed as a tree structure of categories. In this tree, sub-categories represent more specific categories than their parents. For example, the branch "Top/Arts/Television/Networks" is split into two sub categories: "Top/Arts/Television/Networks/Cable" and "Top/Arts/Television/Networks/Satellite". "Cable" is further divided into two sub categories: "HD" and "Movies". Each category contains links to web sites in the form of URL entries. Specifically, the ODP entries are either in the *domain-level* URL form or in the *URL-level* URL form and each is associated with some category C. For example, money.cnn.com is in the *domain-level* URL form and cnn.com/CNN/Programs is in the *URL-level* URL form, and are both associated with category "Arts/Television/Networks/Cable/CNN".

It is important to note, that while the ODP entries are either at the *domain-level* URL or at a more detailed form (*URL-level* URL), the input that we expect is only at the *domain-level* URL form. Thus, we define two distance functions, as follows.

Domain Distance

Given two *domain-level* URLs d_i and d_j, $d_{domain}(d_i, d_j)$ returns the length of the longest common suffix between d_i and d_j. For example, the distance between d_i = a.b.c and d_j = d.b.c is 2. We further normalize the distance by d_i size, that is defined as the number of ("." + 1) in d_i. Hence, in this example, the normalized distance $\widehat{d_{domain}(d_i, d_j)} = 2/3$.

Let us assume that d_i is a URL that corresponds to some domain browsed by some user u and d_j is a domain-level URL that appears in some ODP category C.

---

[2]http://www.dmoz.org/.

Obviously, the largest $\overline{d_{domain}}(d_i, d_j)$ is, the highest the chance that this category C is relevant to u. Specifically, the normalized distance 1 is optimal and happens when there is an exact match between the input URL to some domain-level URL that appears in an ODP category.

For example, assume an input $d_i = \text{a.b.c}$, and the ODP includes URLs a.b.c, d.b.c, e.b.c and x.c. Then, the normalized distances $\overline{d_{domain}}(d_i, d_j)$ are $1, 2/3, 2/3$, and $1/3$, respectively. When there is an exact match between d_i and d_j, and d_j belongs to category C, we simply associate C with d_i. On the other hand, when there is no exact match between d_i and d_j (i.e., $\overline{d_{domain}}(d_i, d_j) < 1$) we still want to associate some category with d_i. Thus, we define that, if the majority of the relevant domain-level URLs that exist in the ODP belong to some category C, this category C will be associated with d_i. In order to do so, we apply a weighted voting approach [14, 17] to each of the relevant domain level URLs d_j, such that its distance, $d = d_{domain}(d_i, d_j) > 1$ votes proportionally to d. If the (weighted) majority agree on a category C (i.e., more than 50% of the total votes), C will be reported as the matching category to the input domain-level URL, d_i. Formally,

$$\textbf{if } \frac{\sum_{\forall d_j \in C_k, \forall d_{domain}(d_i,d_j)>1} \overline{d_{domain}}(d_i, d_j)}{\sum_{\forall d_j, \forall d_{domain}(d_i,d_j)>1} \overline{d_{domain}}(d_i, d_j)} \geq 0.5, \textbf{ then } \quad \{ \text{ associate } C_k \text{ with } d_i\}$$

For example, assume that $d_i = \text{a.b.c}$, and the ODP includes entries d.b.c, e.b.c, f.b.c and x.y.c with the corresponding categories: C_1, C_1, C_2 and C_2. The normalized distances $\overline{d_{domain}}(d_i, d_j)$ are $2/3, 2/3, 2/3$, and $1/3$, respectively. Thus, for category C_1 the total weighted votes are $\frac{\frac{2}{3}+\frac{2}{3}}{\frac{2}{3}+\frac{2}{3}+\frac{2}{3}} = \frac{2}{3}$. Category C_2 has total votes of $\frac{\frac{2}{3}}{\frac{2}{3}+\frac{2}{3}+\frac{2}{3}} = \frac{1}{3}$. Consequently, we assign category C_1 to input d_i.

Path Distance

Given a *domain-level* URL d_i and a *URL-level* URL u_j, such that the domain in u_j is d_i, we define $d_{path}(d_i, u_j)$ to be the length of the URL path in u_j. For example, the distance $d_{path}(d_i, u_j)$ between $d_i = \text{a.b.c}$ and ODP URL level entry $u_j = \text{a.b.c/d}$ is 1. Here, the smaller the distance is, the better is the match. For example, assume the input is $d_i = \text{a.b.c}$, and the ODP includes entries a.b.c/d, a.b.c/e, a.b.c/f and a.b.c/g/h. The first three entries have the same distance $d_{path}(d_i, u_j) = 1$, while the last ODP entry (a.b.c/g/h) has a distance of 2. In such a case, there is no exact match between the inputs, still, we want to associate some category with d_i. Again, we apply the weighted voting approach where each of the ODP entries u_j with $d_{path}(d_i, u_j) \geq 1$ votes inverse proportionally to their distance value. If the (weighted) majority agree on a category, this will be reported as the matching category to the input domain-level URL. Formally,

$$\textbf{if} \frac{\sum_{\forall u_j \in C_k, \forall d_{path}(d_i, u_j) \geq 1} \frac{1}{d_{path}(d_i, u_j)}}{\sum_{\forall u_j, \forall d_{path}(d_i, u_j) \geq 1} \frac{1}{d_{path}(d_i, u_j)}} \geq 0.5, \textbf{then} \quad \{d_i \text{category is } C_k\}$$

For example, assume the input $d_i =$ a.b.c and the ODP includes entries a.b.c/d, a.b.c/e/i, a.b.c/f/j and a.b.c/g/h, with the corresponding categories: C_1, C_2, C_2 and C_2. The first entry's distance is 1, whereas, the rest 3 entries have distance $d_{path}(d_i, u_j) = 2$. Thus, for category C_1 the total weighted votes are $\frac{1}{1 + \frac{1}{2} + \frac{1}{2} + \frac{1}{2}} = \frac{2}{5}$. Category C_2 has total votes of $\frac{\frac{1}{2} + \frac{1}{2} + \frac{1}{2}}{1 + \frac{1}{2} + \frac{1}{2} + \frac{1}{2}} = \frac{3}{5}$. Thus, we assign category C_2 to input d_i.

The categorization process first tries to associate a category to input URLs, using the *Domain distance*. However, if no category is associated at this point (i.e., there was no agreement between the ODP entries about the category) it continues to the *Path distance* calculation. It is worth noting that more complex approaches that combine the different levels, as well as approaches that consider partial ODP paths are possible. We leave this as future work.

In addition, we added some rule-based categorization for those inputs URLs that did not match with any category. We defined 10 very simple rules, such as: if input contains *hotel* or *travel* then categorize it as "Top/Recreation/Travel".

4.3.2.3   Aggregation

Each user is associated with many URLs in the web logs. As a result, reporting each individual's corresponding ODP categories without applying any summarization technique, will make it very difficult, or even impossible, to derive the user behavior and patterns. Thus, the model should support aggregation of the individual entries into some meaningful profile.

Aggregation is inspired by the *GROUP BY* operator in database systems that projects rows having common values into groups. To do so, one needs to define the attributes that will be part of the aggregation, termed as *aggregation level*, followed by the *aggregation function* that evaluates one or more functions on the objects of groups. Typical examples are *sum* or *average* of an attribute in a group. The attributes that are processed by the *aggregation function* are defined by *aggregation selection*.

For example, assume that we want to aggregate for each user, the different categories per day and according to different agent types. We then want to present the duration that the user spent in each of the groups, given an EDR schema, EDR(user_id, url, duration, date, agent_type, ...), an *aggregation level*={user_id, category, date, agent_type}, an *aggregation function* = {sum}, and an *aggregation selection* = {duration}.

4.3.2.4   MapReduce Implementation

MapReduce is a framework for processing large scale data in a distributed fashion. As discussed earlier, due to the massive amount of data that telecom companies have, this computation paradigm is unavoidable. In what follows, we describe the map and reduce functions of the user profile model. Given an EDR record schema, as described in Sect. 4.1.3.2, an *aggregation level*, an *aggregation function* and an *aggregation selection*, we define the mapper and reducer to be:

```
function map(key, EDRrec) {
  category =extractODPCategory(EDRrec.url)
  aggregation_level = aggregation_level ⋃ category
  emit(aggregation_level, aggregation_selection);}

function reduce(aggregation_level, aggregation_selection) {
  s = 0
  foreach rec
    s += aggregation_function(aggregation_selection)
  emit(aggregation_level, [s]);},
```

where the *extractODPCategory(EDRrec.url)* method implements an access to Apache Lucene Index[3] that contains the ODP entries and their associated categories. Thus, fetching the most relevant ODP category for the given EDRrec.url.

For example, assume that we want to aggregate for each user the different categories per day and according to different agent types. We then want to present the duration that the user spent in each of the groups. Given an EDR schema, EDR(user_id, url, duration, date, agent_type, . . .), the mapper and reducer are:

```
function map(key, EDRrec) {
  category =extractODPCategory(EDRrec.url)
  aggregation_level ={EDRrec.user_id,category,
                      EDRrec.date,EDRrec.agent_type}
  emit(aggregation_level, EDRrec.duration);}

function reduce(aggregation_level, EDRrec.duration) {
  s = 0
  foreach rec
    s += sum(EDRrec.duration)
  emit(aggregation_level, [s]);}
```

---

[3]http://lucene.apache.org/.

4.4   System Architecture

Our user profile model is implemented as part of a general analytics library, called *Customer Analyst*. In this section, we describe the Customer Analyst library in general and exemplify its use through the implementation of the user profile model we just presented. We then describe the integration with other analysis tools.

4.4.1   The Customer Analyst Library

Customer Analyst is an analytics library that consists of different reusable components (presented in Fig. 4.1), organized in four layers. The library provides analysts the flexibility to implement various scenarios, by composing those components into complete flows. We will now describe the different layers in the library, focusing on different Telecom scenarios. Note that the library itself is industry agnostic.

At its core, the library needs to support massive-scale data scenarios, and thus, its infrastructure is built on top of Apache Hadoop (in practice embodied into the IBM InfoSphere BigInsights platform[4]). Specifically, it utilizes the Jaql Language,[5] Java MapReduce and the Oozie workflow[6] platform.

4.4.1.1   Utils

The JSON (JavaScript Object Notation)[7] format has become one of the most popular formats for big data: being self-describing and relatively compact, it provides a flexible way to share information between systems. Therefore, Customer Analyst provides methods for efficient parsing, creation and manipulation of JSON format files. Those utilities are part of the *Utils* components. Another important role of the *Utils* components is the *Data Cleansing* capability: the role of those utilities is to standardize the input and removing noise. In the case of *URL* analysis, the cleansing simply includes removing the leading "http://", the port information, etc.

4.4.1.2   Resource Analysis Layer

This layer includes algorithms that receive the raw data (usually after cleansing) as input, and extract different features that will be consumed later by other layers.

---

[4]http://www-01.ibm.com/software/data/infosphere/biginsights/.

[5]http://code.google.com/p/jaql/.

[6]http://rvs.github.com/oozie/index.html.

[7]http://www.json.org/.

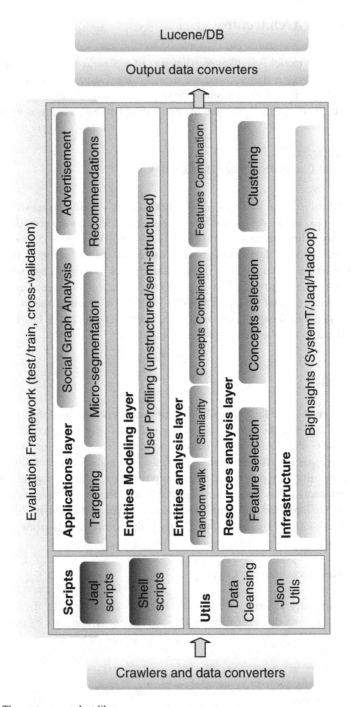

Fig. 4.1 The customer analyst library

Specifically, the library contains three different analysis components, namely *Features Selection*, *Concepts Selection* and *Clustering*.

The *Features Selection* component includes a set of algorithms that aim at selecting a subset of the features (e.g., terms) from a text and eliminating noise. In the literature, there are many approaches to feature selection; the state-of-the-art include Mutual Information, Chi-square [18], and Kullback-Liebler Divergence [24].

Another component in this layer is the *Concepts Selection*, which receives raw textual data or features as input and associates them with concepts. This component harnesses concepts from different taxonomies, including open source taxonomies, such as Wikipedia,[8] and the Open Directory Project (ODP). In Sect. 4.4.2 we have described in details how this is utilized for the purpose of user profiling.

The last component in this layer is the *Clustering* component. It is used to implement scenarios that include clustering of feature vectors, for example, the clustering of different user profiles into micro-segments.

4.4.1.3   Entities Analysis Layer

This layer consumes the output of the *Resource Analysis Layer* and implements algorithms, such as random walk [21] that are relevant for social graph analysis that can be used in churn models 4.1.4.1. In addition, *similarity* algorithms are a fundamental part of any models of targeted advertisement, as described in Sect. 4.1.4.2. Finally, the combination components, namely the *Features Combination* and *Concepts Combination* components, contain implementations of different aggregation methods, including the smoothing parameters (that need to be learned using machine learning techniques) in case of weights sum, etc.

4.4.1.4   Entities Modeling Layer

We have described some basic features and algorithms as part of the analysis layers. The *Entities Modeling Layer* handles the modeling part of the library and includes different user profile models, for example, the user model for the telecom use case we have defined in Sect. 4.3.2. Another supported approach for representing a user profile, based on content consumed or produced by a user, is as a weighted vector of terms, also known as the *Bag of Words* (BOW) model [2, 25]. The work in [24] describes such a user profile model.

---

[8]http://www.wikipedia.org/.

4.4.1.5 Applications Layer

In this layer, are defined the specific applications of the library. For each application, the layer contains the specification of the inputs, outputs, and an implementation of a workflow that utilizes the different components at the lower layers. For example, in our case, we considered inputs such as: distance functions, number of categories, etc. The outputs are the entities from the modeling layers, and the workflow is described below in Sect. 4.4.2.

4.4.1.6 Evaluation Framework

Customer Analyst provides a robust framework for testing and validating results: cross-validation is used to assess the statistical results of the different analyses, including the partitioning of a sample of data into complementary subsets, performing the analysis on one subset (training set) and validating the analysis on the other subset (test set).

Finally, the Customer Analyst library provides an API for inputs and outputs. The library input supports both data push and data pull models and thus integrates the Apache Nutch crawler.[9] In addition, it supports different data conventions (such as encoding). As the Customer Analyst is not a silo, the outputs of the library are usually consumed by other analytics tools, such as SPSS,[10] and thus the output can be saved into databases or textual indices.

4.4.2 Analysis Flow

In this section we present how the Customer Analyst library is used to generate the user profiles. Figure 4.2a shows the schematic flow using the Customer Analyst components, including the inputs and intermediate outputs (in dashed rectangles); such a flow can be implemented using a Oozie workflow, as shown in the Appendix (Listings 4.1). The reader should note the execution order, what can be done in parallel, etc. The flow starts with the ODP data as an input, where the data is represented in RDF format,[11] and goes into the data cleansing component. At this stage, parsing and cleansing of the data is done and the output of this step is indexed into an Apache Lucene Index structure (the ODP index).

In the next step, the concepts selection, the inputs are triplets: the URLs associated with the user id u and other *context* information, such as agent type, content type and date, and the ODP index built at the previous step. To recall, in

---

[9]http://nutch.apache.org/.

[10]http://www-01.ibm.com/software/analytics/spss/.

[11]http://www.w3.org/RDF/.

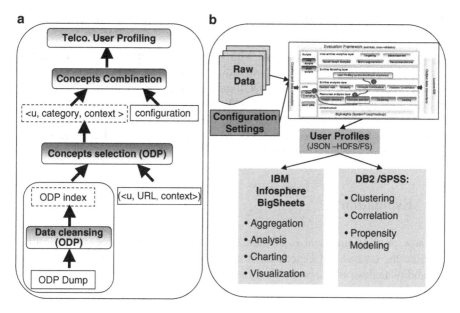

Fig. 4.2 (**a**) Telcom flow using customer analyst, and (**b**) end-to-end flow

Table 4.1 Example of a user profile with top-4 categories

Category	Agent type	Date	Count
Arts/Entertainment/News_and_Media	AndroidBrowser	2011-09-26	22
Arts/Radio/Internet/Directories	AndroidBrowser	2011-09-27	15
Reference/Maps/Google_Maps	BlackberryBrowser	2011-09-27	14
Arts/Entertainment/News_and_Media	AndroidBrowser	2011-09-27	13

this step, the task at hand is to match a category from the ODP taxonomy into the given domain URLs. Thus, the output is a set of triplets $\langle u, d, context \rangle$. Those triplets are now consumed by the concepts combination component together with some configuration parameters, such as how to aggregate the concepts, how many categories to report per user and more. Some examples are: "aggregate by user id, category and agent type", "aggregate by user id, category and date", etc. As for aggregation selection, methods like "count" – how many times a category C appears in the user entries' or "duration" – how long this user was browsing web pages that belong to category C", and more, can be used. The output after this combination is the user profile. For example, Table 4.1 shows the top-4 categories for a specific user, aggregated by category, agent type and date using count.

For this user profile, we can understand the different interests, e.g., preferring web sites about "News and Media" (regularly visiting these sites using Android), along with the intensity.

4.4.3 End-to-End Flow

Figure 4.2b depicts a complete flow with some options for analysis based on the generated user profiles. Here, the flow shows the raw data, and configuration files that are inputs to the Customer Analyst library. Then the output, the user profiles can be consumed by the IBM Infosphere BigSheets tool to explore and visualize data,[12] or by analysis tools such as SPSS (on top of IBM DB2). For example, clustering or propensity models are created with SPSS, whereas simple aggregation, visualization and charting may be done with BigSheets.

4.5 Evaluation

We now provide empirical evaluation of our user profiling framework. We first describe the dataset and experimental setting followed by quality evaluation and scalability analysis of the proposed framework.

4.5.1 Dataset and Experimental Setup

For the experimental evaluation we used a real-world dataset containing HTTP session logs, URL, date, time, agent type, content type, and more. The logs cover the browsing history for 30 days and for more than eight millions telecom customers. On average, there are around 30 million entries (browsing requests) per hour. The total volume of data is around 5 TB.

Experiments were conducted using a 8-nodes commodity machines cluster (each machine with 2 GB RAM, 4 cores), each node with Red Hat Linux operating system running IBM Infopshere BigInsights, which contains Hadoop version 0.20.1.

4.5.2 Quality Analysis

We start with quality analysis of our proposed categorization approach and the user profile model, as described in Sect. 4.3.2.

4.5.2.1 ODP Categorization

We compare the ODP-based categorization presented above with a state-of-the-art web filtering product, include spam detection and content filtering. Specifically,

---

[12]http://www-01.ibm.com/software/ebusiness/jstart/bigsheets/.

Table 4.2 Top-10 categories along with their ODP precision and recall values

Category	Recall	Precision
Social networking	0.986	0.995
Software/hardware	0.965	1
Banner advertisements	0.731	0.963
Search engines/web catalogues/portals	0.924	0.869
General business	0.328	0.916
News/magazines	0.916	0.970
Education	0.842	1
Auctions/classified Ads	1	0.833
Cinema/television	0.888	1

this product classifies web pages based on the layout, text, hypelinks, etc., and offers more than 14 million URLs conveniently categorized into 60 different content groups (such as "Social Networking", "News/Magazines" and more).

The prominent difference between this product and our approach is the granularity of the categories. This is due to the different needs of the two tools. While for web filtering applications 60 categories are enough to classify web pages, for user profiling tasks this is coarse and not descriptive enough. For example, http://data.flurry.com is categorized as "Software/Hardware" by the filtering product, whereas the associated ODP category is "Top/Business/Telecommunications/Software", which is much more descriptive. This allows the analysts to gain a better insights and understanding about the customer interests, in this example, understanding that the customer is interested in telecommunication software and not in any general software.

To study the quality of the ODP categories, we extracted 1,000 URLs (Uniform at Random) from the logs. Each URL was categorized by both the filtering product and using our approach. We define the filtering product categorization as the "ground truth", and manually assess each of the ODP categories with respect to the ground truth category, measured by the precision and recall.

Table 4.2 summarizes the precision and recall values for the top-10 categories. In general, the recall and precision of the top-10 categories are very good, the average precision and recall values are 0.95 and 0.84, respectively. A low recall score was obtained for category "General Business"; this was due to the fact that many missing URLs in the ODP were classified as "General Business" in the filtering product.

4.5.2.2 ODP Coverage

As discussed in Sect. 4.3.2.2, there is a gap between the input form (domain level) and the ODP entries (URL and domain forms), which is solved by introducing the different distance functions (*Domain Distance* and *Path Distance*). In total, we were able to categorize 87% of the input (the rest 13% were "Null").

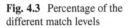 **Fig. 4.3** Percentage of the
different match levels

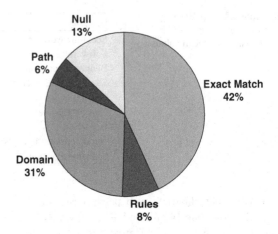

Figure 4.3 shows the precentage of the input that was categorized by the different approaches. For example, 42% of the input had exact match with ODP entries, 31% of the input was categorized using the *Domain Distance* and only 6% was categorized with the *Path Distance* approach.

4.5.3 Scalability Analysis

We now analyze the scalability of our framework with respect to several parameter settings. The main factor that affects the performance of the flow is the data size, i.e., the number of entries (HTTP sessions) that need to be processed.

4.5.3.1 Effect of Data Size

We first analyze the performance with respect to the data size. It is worth noting that instead of looking on data size directly, we looked on a slightly different parameter which is the number of processed days. The HTTP logs were partitioned into "hourly" logs, i.e., there are 24 logs per day. Thus, is seems natural to choose this unit (i.e., day) for performance analysis. As mentioned above, in our dataset there are, on average, 30 million browsing requests per hour. We vary the number of processed days from 1 to 25 days (in increments of 1, 5, 10 and 25 days) and measure the runtime. Figure 4.4 provides the runtime for the different number of days. We observe that the runtime increases linearly with the increment in the number of processed days. For example, analyzing 1 day data took 17 min and analyzing 10 days data took 157 min.

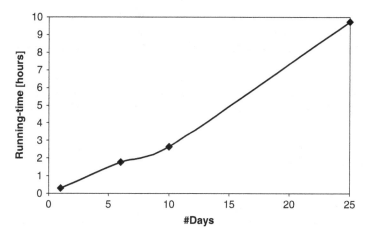

Fig. 4.4 Runtime performance (in hours) with respect to increasing in number of processed days

Table 4.3 Increase in runtime with respect to different aggregation levels

Aggregation level	Increase in runtime (%)
{date,content_type}	11
{date,content_type,user_agent}	24
{date,content_type,user_agent,URL}	53
{date,content_type,user_agent,URL,http_method}	58

4.5.3.2   Effect of Aggregation

Finally, we analyze the scalability of our framework with respect to the *aggregation level*. Recall that the *aggregation level* refers to the way that the entries per users are grouped together (see Sect. 4.3.2.3). We define the basis aggregation level to be by the "date" attribute and then compare the different levels with respect to this basis level.

Table 4.3 summarizes the percentage of increase in running time for the different levels of aggregation per user. As expected, the runtime increases when we add more levels of aggregation. This is due to the fact that more outputs need to be written by both mappers and reducers. For instance, when we aggregate using $\{date, content\_type, user\_agent, URL\}$, the runtime is 50% more than the runtime of the basis aggregation ("date"). It is worth noting that in our implementation the different *aggregation functions* did not affect the runtime.

4.6   Conclusions

In this chapter we presented a scalable user profiling solution for the telecom industry, implemented on top of the Apache Hadoop framework. We presented the general architecture and a complete scenario, including the implementation details.

Future work includes the extension of our framework with other paradigms, such as hierarchical or semantic models. We also intend to incorporate structured data sources into our framework.

Acknowledgements The authors would like to thank Shai Erera and Gilad Barkai for the useful discussions about implementation issues. We also thank Haggai Roitman for sharing thoughts and ideas. Finally, we thank Matin Jouzdani for his support to make it a successful project.

Appendix

Listing 4.1 Customer Analyst's Oozie workflow

```xml
<workflow-app xmlns='uri:oozie:workflow:0.1' name='map-reduce-wf'>
    <start to='ODPcleansingJob'/>
    <action name='ODPcleansingJob'>
        <map-reduce>
            <job-tracker>\${jobTracker}</job-tracker>
            <name-node>\${nameNode}</name-node>
            <configuration>
                <property>
                    <name>mapred.job.name</name>
                    <value>ODP-cleansing-indexing</value>
                </property>
                <property>
                    <name>mapreduce.map.class</name>
                    <value>com.ibm.analyze.mr.ODPMapper</value>
                </property>
                <property>
                    <name>mapreduce.reduce.class</name>
                    <value>com.ibm.analyze.mr.ODPReducer</value>
                </property>
                <property>
                    <name>mapred.reduce.tasks</name>
                    <value>1</value>
                </property>
                <property>
                    <name>mapred.input.dir</name>
                    <value>\${inputDir}</value>
                </property>
                <property>
                    <name>mapred.output.dir</name>
                    <value>/usr/output/\${wf:id()}/ODPindex</value>
                </property>
            </configuration>
        </map-reduce>
    </action>
    <ok to="fork"/>
    <error to="kill"/>
    <fork name='fork'>
        <path start='conceptSelectionODP' />
    </fork>
    <action name="conceptSelectionODP">
        <map-reduce>
            <job-tracker>\${jobtracker}</job-tracker>
            <name-node>\${namenode}</name-node>
            <configuration>
                <property>
                    <name>mapred.job.name</name>
```

```xml
                            <value>concept-selection-ODP</value>
                        </property>
                        <property>
                            <name>mapred.mapper.class</name>
                            <value>com.ibm.analyze.mr.ConceptSelectMapper</value>
                        </property>
                        <property>
                            <name>mapred.reducer.class</name>
                            <value>com.ibm.analyze.mr.ConceptSelectReducer</value>
                        </property>
                        <property>
                            <name>mapred.input.dir</name>
                            <value>/usr/output/\${wf:id()}/ODPindex</value>
                        </property>
                        <property>
                            <name>mapred.output.dir</name>
                            <value>/usr/output/\${wf:id()}/UserConcepts</value>
                        </property>
                    </configuration>
                </map-reduce>
            </action>
            <ok to="join"/>
            <error to="kill"/>
            <action name="conceptCombinationODP">
                <map-reduce>
                    <job-tracker>\${jobtracker}</job-tracker>
                    <name-node>\${namenode}</name-node>
                    <configuration>
                        <property>
                            <name>mapred.job.name</name>
                            <value>concept-combination-ODP</value>
                        </property>
                        <property>
                            <name>mapred.mapper.class</name>
                            <value>com.ibm.analyze.mr.ConceptCombineMapper</value>
                        </property>
                        <property>
                            <name>mapred.reducer.class</name>
                            <value>com.ibm.analyze.mr.ConceptCombineReducer</value>
                        </property>
                        <property>
                            <name>mapred.input.dir</name>
                            <value>/usr/output/\${wf:id()}/ODPindex,
                                   /usr/output/\${wf:id()}/UserConcepts</value>
                        </property>
                        <property>
                            <name>mapred.output.dir</name>
                            <value>\${output}</value>
                        </property>
                        <property>
                            <name>schema</name>
                            <value>\${schema}</value>
                        </property>
                        <property>
                            <name>rank</name>
                            <value>\${rank}</value>
                        </property>
                        <property>
                            <name>aggregateFields</name>
                            <value>\${aggregateFields}</value>
                        </property>
                    </configuration>
                </map-reduce>
                <ok to="end"/>
                <ok to="kill"/>
            </action>
            <kill name="kill">
```

```
      <message>[\${wf:errorMessage(wf:lastErrorNode())}]</message>
   </kill>
   <end name='end'/>
</workflow-app>
```

References

1. Cetintemel, U., Franklin, M.J., Giles, C.L.: Self-adaptive user profiles for large-scale data delivery. In: ICDE, San Diego, pp. 622–633 (2000)
2. Chen, L., Sycara, K.: Webmate: a personal agent for browsing and searching. In: AGENTS '98, St. Paul. ACM, New York (1998)
3. Chen, Y., Pavlov, D., Canny, J.F.: Large-scale behavioral targeting. In: KDD '09, Paris. ACM, New York (2009)
4. Chirita, P.A., Nejdl, W., Paiu, R., Kohlschütter, C.: Using ODP metadata to personalize search. In: Proceedings of the 28th Annual International ACM SIGIR Conference on Research and Development in Information Retrieval, SIGIR '05, Salvador, pp. 178–185. ACM, New York (2005)
5. Cohn, D., Hofmann, T.: The missing link – a probabilistic model of document content and hypertext connectivity. In: Advances in Neural Information Processing Systems, Vancouver (2001)
6. Davidov, D., Gabrilovich, E., Markovitch, S.: Parameterized generation of labeled datasets for text categorization based on a hierarchical directory. In: SIGIR '04, Sheffield (2004)
7. Dean, J., Ghemawat, S.: Mapreduce: simplified data processing on large clusters. Commun. ACM 51(1), 107–113 (2008)
8. de Reuver, M., Haaker, T.: Designing viable business models for context-aware mobile services. Telemat. Inform. 26(3), 240–248 (2009)
9. Ghemawat, S., Gobioff, H., Leung, S.-T.: The Google file system. In: Proceedings of the Nineteenth ACM Symposium on Operating Systems Principles, SOSP '03, Bolton Landing, pp. 29–43. ACM, New York (2003)
10. Gyöngyi, Z., Garcia-Molina, H., Pedersen, J.: Combating web spam with trustrank. In: Proceedings of the Thirtieth International Conference on Very Large Data Bases – Volume 30, VLDB '04, Toronto, pp. 576–587 (2004)
11. http://en.wikipedia.org/wiki/Privacy_internet
12. https://www.eff.org/deeplinks/2012/03/best-practices-respect-mobile-user-billrights
13. Hung, S.-Y., Yen, D.C., Wang, H.-Y.: Applying data mining to telecom churn management. Expert Syst. Appl. 31, 515–524 (2006)
14. Ingrid, D.: Weighted voting systems. Voting and Social Choice (2002)
15. Kaasinen, E.: User needs for location-aware mobile services. Pers. Ubiquitous Comput. 7, 70–79 (2003)
16. Kleinberg, J.M.: Authoritative sources in a hyperlinked environment. J. ACM 46(5), 604–632 (1999)
17. Larry, B.: Weighted Voting Systems (2001)
18. Manning, C.D., Raghavan, P., Schutze, H.: Introduction to Information Retrieval. Cambridge University Press, New York (2008)
19. Nunes, M., Cabral, L., Lima, R., Freitas, F., Reinaldo, G., Prudencio, R.: Docs-Clustering: A System for Hierarchical Clustering and Document Labeling (2008)
20. Oishi, T., Kambara, Y., Mine, T., Hasegawa, R., Fujita, H., Koshimura, M.: Personalized search using ODP-based user profiles created from user bookmark. In: PRICAI 2008: Trends in Artificial Intelligence, Hanoi. Volume 5351 of Lecture Notes in Computer Science, pp. 839–848 (2008)
21. Pearson, K.: The problem of the random walk. Nature 72, 294 (1905)

22. Qi, X., Davison, B.D.: Web page classification: features and algorithms. ACM Comput. Surv. **41**(2), 1–31 (2009)
23. Richter, Y., Yom-Tov, E., Slonim, N.: Predicting customer churn in mobile networks through analysis of social groups. In: SDM, Columbus (2010)
24. Shmueli-Scheuer, M., Roitman, H., Carmel, D., Mass, Y., Konopnicki, D.: Extracting user profiles from large scale data. In: MDAC, Raleigh (2010)
25. Sugiyama, K., Hatano, K., Yoshikawa, M.: Adaptive web search based on user profile constructed without any effort from users. In: WWW, Manhattan, pp. 675–684 (2004)
26. Tanudjaja, F., Mui, L.: Persona: a contextualized and personalized web search. In: Proceedings of the 35th Annual Hawaii International Conference on System Sciences, Big Island, p. 67 (2001)
27. van Setten, M., Pokraev, S., Koolwaaij, J.: Context-aware recommendations in the mobile tourist application compass. In: Adaptive Hypermedia and Adaptive Web-Based Systems, Eindhoven, vol. 3137, pp. 515–548 (2004)
28. Williamson, M.: Using DMOZ open directory project lists with novell bordermanager (2003)
29. Zhou, Y., Wilkinson, D., Schreiber, R., Pan, R.: Large-scale parallel collaborative filtering for the netflix prize. In: AAIM '08, Shanghai, pp. 337–348. Springer, Berlin/Heidelberg (2008)

Chapter 5
Machine Learning Algorithm Acceleration Using Hybrid (CPU-MPP) MapReduce Clusters

Sergio Herrero-Lopez and John R. Williams

Abstract The uninterrupted growth of information repositories has progressively led data-intensive applications, such as MapReduce-based systems, to the mainstream. The MapReduce paradigm has frequently proven to be a simple yet flexible and scalable technique to distribute algorithms across thousands of nodes and petabytes of information. Under these circumstances, classic data mining algorithms have been adapted to this model, in order to run in production environments. Unfortunately, the high latency nature of this architecture has relegated the applicability of these algorithms to batch-processing scenarios. In spite of this shortcoming, the emergence of massively threaded shared-memory multiprocessors, such as Graphics Processing Units (GPU), on the commodity computing market has enabled these algorithms to be executed orders of magnitude faster, while keeping the same MapReduce-based model. In this chapter, we propose the integration of massively threaded shared-memory multiprocessors into MapReduce-based clusters, creating a unified heterogeneous architecture that enables executing Map and Reduce operators on thousands of threads across multiple GPU devices and nodes, while maintaining the built-in reliability of the baseline system. For this purpose, we created a programming model that facilitates the collaboration of multiple CPU cores and multiple GPU devices towards the resolution of a data intensive problem. In order to prove the potential of this hybrid system, we take a popular NP-hard supervised learning algorithm, the Support Vector Machine (SVM), and show that a $36 \times -192 \times$ speedup can be achieved on large datasets without changing the model or leaving the commodity hardware paradigm.

S. Herrero-Lopez (✉)
Technologies, Equities and Currency (TEC) Division, SwissQuant Group AG, Kuttelgasse 7, 8001 Zurich, Switzerland
e-mail: sergherrero@gmail.com

J.R. Williams
Massachusetts Institute of Technology, 77 Massachusetts Avenue, 02139 Cambridge, MA, USA
e-mail: jrw@mit.edu

A. Gkoulalas-Divanis and A. Labbi (eds.), *Large-Scale Data Analytics*,
DOI 10.1007/978-1-4614-9242-9_5, © Springer Science+Business Media New York 2014

5.1 Introduction

The data mining community has often assumed that performance increase on existing techniques would be given by the continuous improvement of processor technology. Unfortunately, due to physical and economic limitations, it is not recommendable to rely on the exponential frequency scaling of CPUs anymore. Furthermore, the low price and ubiquity of data generation devices not only has led to larger datasets that need to be digested on a timely manner, but also to the growth of dimensionality, categories and formats of the data. Simultaneously, an increasingly heterogeneous computing ecosystem has defined three computing families:

1. *Commodity Computing*: It encompasses large-scale geographically distributed commodity machine clusters running primarily open source software. Its reliability to host batch processing systems, such as Hadoop [12], and storage systems, such as BigTable [5] or Cassandra [17], across tens of thousands of nodes and petabytes of data, have made commodity computing the foundation of internet-scale companies and the cloud.
2. *High Performance Computing/Supercomputing*: It refers to centralized multi-million computer systems capable of delivering high throughput for complex tasks that demand large computational power. Typically, these are funded and operated by governments or large corporations, and are utilized for the resolution of scientific problems.
3. *Appliance Computing*: It refers to highly specialized systems exclusively designed to carry out one or few similar tasks with maximum performance and reliability. These nodes combine state-of-the-art processor, storage and interconnect technologies and cost one order of magnitude less than supercomputers. These computing appliances have been successfully utilized for large-scale analytics and enterprise business intelligence operations.

Under these circumstances, it is required for the research community to investigate the adaptation of classic and novel data intensive algorithms to this heterogeneous variety of parallel computing ecosystems and the technologies that compose them. This adaptation process can be separated into two phases: The *Extraction Phase*, in which the parallelizable parts, called parallel *Tasks*, of the algorithm are identified and separated; and the *Integration Phase*, in which these tasks are implemented for the most suitable parallel computing platform or combination of them.

5.1.1 Extraction Phase

The *extraction of parallelism* on a data intensive algorithm can be carried out at different levels, with different impacts on performance and increased programming complexity:

1. *Independent Runs:* This is the most common technique; it simply runs the same algorithm with different configuration parameters on different processing nodes. Each of the runs is independent and parallel execution does not speed up individual runs.
2. *Statistical Query & Summation:* This technique decomposes the algorithm into an adaptive sequence of statistical queries, and parallelizes these queries over the sample [16]. This approach is satisfactory in speeding up slow algorithms, in which little communication is needed.
3. *Structural Parallelism:* This technique is based on the exploitation of fine-grained data parallelism [15]. This is achieved by handling each data point with one or few processing threads.

These three techniques are complementary and are often combined to yield maximum performance on a given target parallel computing platform. Successful parallelization transforms a computationally limited problem into a bandwidth bound problem, in which communication between processing units becomes the bottleneck and optimizing for minimum latency gains critical importance. The full exposure of the complexity of parallel programming will result in the largest performance gain.

Individual parallel tasks extracted through both Statistical Queries & Summation and/or Structural Parallelism, can be directly modeled using the MapReduce programming paradigm [8]. The MapReduce framework is illustrated in Fig. 5.1.

The Map and Reduce operators are defined with respect to structured (key, value) pairs. Map (M) takes one pair of data with a type in one domain, and returns a list of pairs in a different domain:

$$M [k_1, v_1] \rightarrow [k_2, v_2] \tag{5.1}$$

The Map operator is applied in parallel to every item in the input dataset. This produces a list of (k_2, v_2) pairs for each call. Then, the framework collects all the pairs with the same key and groups them together. The Reduce (R) operator is then applied to produce a v_3 value.

$$R [k_2, \{v_2\}] \rightarrow [v_3] \tag{5.2}$$

The advantage of the MapReduce model is that makes parallelism explicit, and more importantly, language or platform agnostic, which allows executing a given algorithm on any combination of platforms in the parallel computing ecosystem. M or R tasks are distributed dynamically among a collection of *Workers*. The Workers is an abstraction that can represent nodes, processors or Massively Parallel Processor (MPP) devices.

Researchers have focused their effort on the decomposition of Machine Learning algorithms as iterative flows of Map and Reduce tasks. Next, the decomposition of three classic Machine Learning algorithms into flows of Map (M) and Reduce (R) tasks is explained:

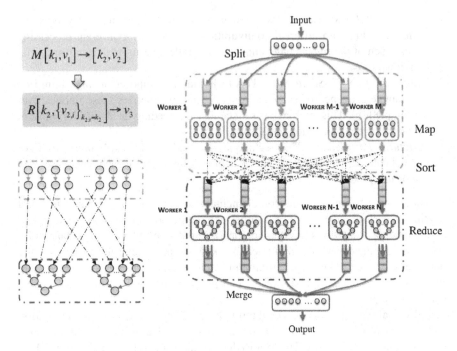

Fig. 5.1 MapReduce primitives and runtime

- *K-means*: The *K*-means clustering algorithm can be represented as an iterative sequence of (M, R) tasks that run until the stop criteria are met. M represents the assignation of points to clusters and R the recalculation of the cluster centroids. *K*-means is illustrated in Fig. 5.2.
- *Expectation Maximization*: The EM algorithm for Gaussian mixtures is represented by iterations of (M, R, R, R) tasks running until convergence. M corresponds to the E-step of the algorithm, while (R, R, R) correspond to the M-step that calculates the mixture weights a_i, means $\bar{\mu}_k$ and covariance matrices Σ_k, respectively. EM is illustrated in Fig. 5.3.
- *Support Vector Machine*: The resolution of the dual SVM problem using the Sequential Minimal Optimization (SMO) [20] is represented by iterations of (M, M, R, M) tasks running until convergence. These tasks reproduce the identification of the two Lagrange multipliers to be optimized in each iteration, and their analytic calculation. The SVM is illustrated in Fig. 5.4.

5.1.2   Integration Phase

The *integration* of parallel MapReduce tasks into diverse computing platforms spans a wide and heterogeneous variety of parallel system architectures. Originally, internet-scale companies decomposed indexing and log-processing jobs into Map

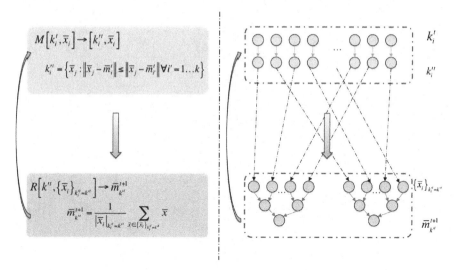

Fig. 5.2 Decomposition of K-means into MapReduce tasks

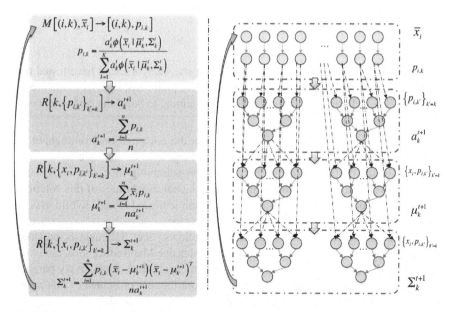

Fig. 5.3 Decomposition of EM using Gaussian mixtures into MapReduce tasks

and Reduce tasks that were executed in batches on top of a distributed file system hosted by hundreds or thousands of commodity nodes. Its proven reliability in production, along with its symbiosis towards virtualized environments, led the MapReduce model to be one of the key data processing paradigms of cloud service infrastructures. Research initiatives have investigated the applicability of

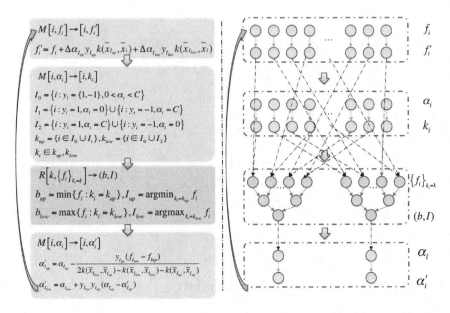

$$M[i,f_i] \rightarrow [i,f_i']$$

$$f_i' = f_i + \Delta\alpha_{I_{up}} y_{I_{up}} k(x_{I_{up}}, x_i) + \Delta\alpha_{I_{low}} y_{I_{low}} k(x_{I_{low}}, x_i)$$

$$M[i,\alpha_i] \rightarrow [i,k_i]$$

$$I_0 = \{i : y_i = \{1,-1\}, 0 < \alpha_i < C\}$$
$$I_1 = \{i : y_i = 1, \alpha_i = 0\} \cup \{i : y_i = -1, \alpha_i = C\}$$
$$I_2 = \{i : y_i = 1, \alpha_i = C\} \cup \{i : y_i = -1, \alpha_i = 0\}$$
$$k_{up} = \{i \in I_0 \cup I_1\}, k_{low} = \{i \in I_0 \cup I_2\}$$
$$k_i \in k_{up}, k_{low}$$

$$R[k, \{f_i\}_{k_i = k}] \rightarrow (b, I)$$

$$b_{up} = \min\{f_i : k_i = k_{up}\}, I_{up} = \text{argmin}_{k_i = k_{up}} f_i$$
$$b_{low} = \max\{f_i : k_i = k_{low}\}, I_{low} = \text{argmax}_{k_i = k_{low}} f_i$$

$$M[i,\alpha_i] \rightarrow [i,\alpha_i']$$

$$\alpha_{I_{up}}' = \alpha_{I_{up}} - \frac{y_{I_{up}} (f_{I_{low}} - f_{I_{up}})}{2k(\bar{x}_{I_{up}}, \bar{x}_{I_{low}}) - k(\bar{x}_{I_{up}}, \bar{x}_{I_{up}}) - k(\bar{x}_{I_{low}}, \bar{x}_{I_{low}})}$$
$$\alpha_{I_{low}}' = \alpha_{I_{low}} + y_{I_{low}} y_{I_{up}} (\alpha_{I_{up}} - \alpha_{I_{up}}')$$

f_i

f_i'

α_i

k_i

$\{f_i\}_{k_i = k}$

(b, I)

α_i

α_i'

Fig. 5.4 Decomposition of SVM into MapReduce tasks

this model in scientific environments and enterprise analytics, and have tested the implementation of MapReduce tasks on alternative platforms, such as multicore or MPPs (GPUs, Cell microprocessors or FPGAs), aiming to boost the performance of computationally expensive jobs.

In this chapter, a hybrid solution that boosts the computational throughput of commodity nodes is proposed, based on the integration of multiple MPPs into the MapReduce runtime. For this purpose, a programming model to orchestrate MPPs is developed. In order to test the computational capabilities of this solution, a multiclass Support Vector Machine (SVM) is implemented for this hybrid system and its performance results for large datasets reported.

The rest of this chapter is organized as follows: Sect. 5.2 reviews previous initiatives that accelerate the execution of data intensive MapReduce jobs, either by optimizing the cluster runtimes or exploiting the capabilities of massively parallel platforms. Section 5.3 enumerates the research contributions presented by this work. Our proposed unified heterogeneous architecture, is described in Sect. 5.4. The decomposition of the SVM problem into MapReduce tasks and its integration into the GPU cluster architecture is explained in Sect. 5.5. Section 5.6 contains details of the performance gain provided by our massively threaded implementation. The conclusions of this work are presented in Sect. 5.7.

5.2   Related Work

In general, the research efforts for the performance improvement of large-scale Machine Learning algorithms, expressed as MapReduce jobs, can be classified into two families, namely *cluster category* and *multiprocessor category*.

Cluster Category: These efforts focus on the adaptation of the cluster runtime to satisfy the particular needs of Machine Learning algorithms and facilitate their integration into clusters. These needs have been identified as: (1) support for iterative jobs, (2) static and bariable data types, and (3) dense and sparse BLAS operations. Ekanayake et al. [10] presented *Twister*, a modified runtime that accommodates multiple Machine Learning algorithms by supporting long-running stateful tasks. Ghoting et al. [11] designed *System-ML*, a declarative language to express Machine Learning primitives and simplify direct integration into clusters. The Apache *Mahout* [1] project compiled a library of the most popular MR-able algorithms for the standard Hadoop implementation of MapReduce.

Multiprocessor Category: In this category MapReduce jobs are scattered and gathered among multiple processing cores on a shared-memory multiprocessor device or multiple devices hosted by interconnected nodes. Typically the processing units in these systems are constructed to run tens of threads simultaneously reducing the load of MapReduce tasks assigned to each thread, while increasing the degree of parallelism. Communication between cores is carried out through the shared-memory hierarchy. Popular systems of this category are Phoenix (multicore) [27], Mars (GPU) [13], CellMR (Cell) [21] and GPMR [22].

The hybrid MapReduce runtime proposed in this chapter is unique in the sense that it combines the best of both worlds to deliver an efficient framework that meets the specific needs of Machine Learning algorithms, and produces up to two orders of magnitude of acceleration using massively threaded hardware.

Particularly for the case of SVM implementations for shared-memory multiprocessors, Chu et al. [23] provide an SVM solver for multicore based on MapReduce jobs obtained through Statistical Query & Summation. Similarly, researchers have been focused on the GPU adaptation of dual form SVMs for binary and multiclass classification [3, 14]. Specifically for the case of SVM implementations in clusters, Chang et al. [6] provide the performance and scalability analysis of a deployment of their PSVM algorithm on Google's MapReduce infrastructure.

5.3   Research Contributions

Typically, both system categories, *Cluster* and *Multiprocessor*, have shown complementary characteristics. Batch processing systems running on commodity clusters provide high reliability through redundancy and near-linear scalability by adding nodes to the cluster for a low cost. Nevertheless, by nature, its intensive cross-machine communication leads to higher latencies and increased complexity for computer cluster administrators. On the contrary, shared-memory multiprocessors do not have any built-in reliability mechanism and their scalability is limited by the number of processing cores and the capacity of the memory hierarchy in place. In these devices, latencies in cross-processor communication are orders of magnitude lower, and single-node execution drastically reduces the system administration complexity. Ideally, a unified system including the benefits of both solutions and meeting the needs of Machine Learning algorithms is desired, in order to execute these algorithms on large scale datasets and obtain the results on a timely manner.

The authors of this chapter believe that both categories can be merged to create a unified heterogeneous MapReduce framework and increase the computational throughput of individual nodes. The contributions of this new hybrid system are the following:

1. *Runtime Adaptation*: The original MapReduce runtime was not designed specifically for Machine Learning algorithms. Even though libraries, such as Mahout [1], implement a variety of classic algorithms, the framework has inherent inefficiencies that prevent it from providing timely responses. Our proposed hybrid design integrates a series of modifications to accommodate some of the common needs of Machine Learning algorithms. These runtime modifications are introduced next:

   - *Iterative MapReduce Jobs*: Most Machine Learning algorithms are iterative. The state of the algorithm is maintained through iterations and is reutilized towards the resolution of the problem in each step. Like *Twister* [10], our solution enables executing long-running iterative jobs that keep a state between iterations.
   - *Static and Variable Data Support*: Most iterative Machine Learning algorithms define two types of data: *static* and *variable*. Static data is read-only and is utilized in every iteration, while variable data can be modified and is typically of smaller size. In order to minimize data movements and memory transfers, our runtime allows specifying the nature of the data.
   - *Dense & Sparse BLAS*: The execution of a task may require as input the results of a dense or sparse BLAS operation. Our solution enables interleaving massively threaded BLAS operations to prepare the input data of M and R steps.

2. *MPP Integration*: As opposed to Mars [13], Phoenix [27] and CellMR [21], which were constructed to run MapReduce jobs within a single isolated multiprocessor and not designed to scale out, our solution takes a different approach based on the integration of MPPs into the existing MapReduce framework as coprocessors. GPMR [22] follows the same direction, but keeping the same runtime and not optimizing it for Machine Learning algorithms.

3. *MPP Orchestration*: A programming model to manage multiple MPPs towards the execution of MapReduce tasks is presented. We use an abstraction, called *Asynchronous Port-based Programming*, which allows creating coordination primitives, such as Scatter-Gather.

4. *Massively Threaded SVM*: While implementations of SVM solvers for multiprocessors and clusters provided satisfactory performance as part of isolated experiments, to the best of our knowledge, this work pioneers the execution of a multiclass SVM on a topology of multiple MPPs intertwining tens of CPU threads and thousands MPP threads collaboratively towards an even faster resolution of the SVM training problem.

5.4   A Unified Heterogeneous Architecture

In this section we provide an overview of the foundations of MapReduce-based batch processing systems. We take the Hadoop architecture as a reference due to its popularity and public nature. First, we explain the characteristics and principles of operation of currently existing data processing nodes, called *Data Nodes (DN)*. Then, we proceed to introduce our modifications by integrating more powerful nodes composed by multiple Massively Parallel Processor (MPP) devices; we call these nodes *MPP Nodes* (MPPN). DNs and MPPNs may coexist within a MapReduce cluster, nevertheless, they are meant to address MapReduce jobs with different requirements: DNs should work on batch, high latency jobs, whereas MPPNs would take responsibility of compute intensive jobs.

5.4.1   MapReduce Architecture Background

Typically, the architecture of MapReduce and MapReduce-like systems consists of two layers: (i) a data storage layer in the form of a Distributed File System (DFS) responsible of providing scalability to the system and reliability through replication of the files, and (ii) a data processing layer in the form of a MapReduce Framework (MRF) responsible of distributing and load balancing tasks across nodes. Files in the DFS are broken into blocks of fixed size and distributed among the DNs in the cluster. The distribution and load balancing is managed centrally in a node called *NameNode* (NN). The NN does not only contain metadata about the files in the DFS, but also manages the replication policy. The MRF follows a master-slave paradigm. There is a single master, called *JobTracker*, and multiple slaves, called *TaskTrackers*. The JobTracker is responsible of scheduling MapReduce jobs in the cluster, along with maintaining information about each TaskTracker's status and task load. Each job is decomposed into MapReduce tasks that are assigned to different TaskTrackers based on locality of the data and their status. In general, the output of the Map task is materialized to the disk before proceeding to the Reduce task. The Reduce task may get shuffled input data from different DNs. Periodically, TaskTrackers sent a heartbeat to the JobTracker to keep it up to date. Typically, TaskTrackers are single or dual threaded and consequently, can launch one or two Map or Reduce tasks simultaneously. Hence, each task is single-threaded and work on a single block point by point sequentially. The architecture is illustrated in Fig. 5.5.

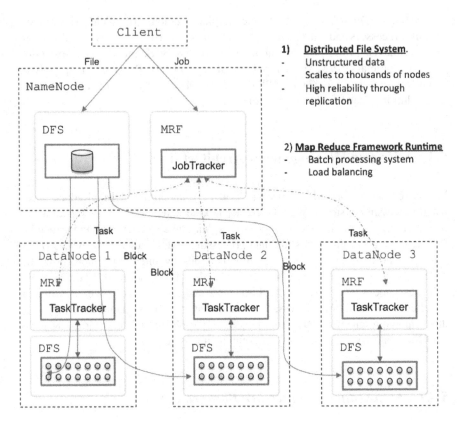

Fig. 5.5 The MapReduce architecture

5.4.2 MPP Integration

We propose the addition of massively parallel processors in order to increase the computational capabilities of the DNs. Currently, DNs use a single thread to process the entire set of data points confined on a given block. This is shown in Fig. 5.6. Parallelism is achieved through the partitioning of data into blocks and the concurrent execution of tasks on different nodes, nevertheless, this setup does not leverage fine-grained parallelism, which can be predominant on data mining algorithms. Fortunately, the introduction of MPPs enables MapReduce tasks to be carried out by hundreds or thousands of threads, giving to each thread one or few data points to work with. This is described in Fig. 5.7. The main differences between DNs and MPPNs are the following:

- *Multithreading*: In DNs the TaskTracker assigns the pair *(Task, Block)* to a single core. Then, the thread running on that core executes the MapReduce function point by point in the block sequentially. On the contrary, in MPPNs the

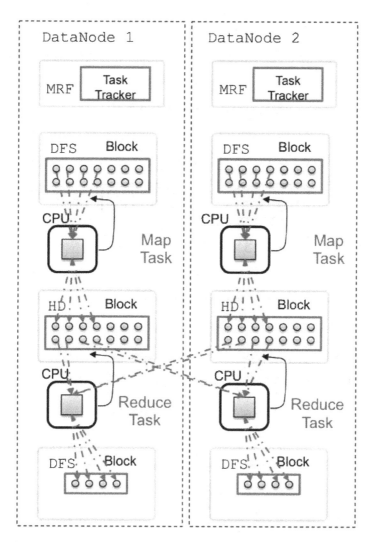

Fig. 5.6 Data node (DN)

TaskTracker assigns the pair *(Task, Block)* to a massively threaded multiprocessor device. Then the device launches simultaneously hundreds or thousands of threads that will execute the same task on multiple data points simultaneously.

- *Pipelining*: In DNs the intermediate result generated by the Map task is materialized by writing the result locally in the node. Before the execution of the Reduce task, the intermediate result is read from the disk and possibly transmitted over the network to a different DN as part of the shuffling process. On the contrary, MPPNs do not materialize the intermediate result. The output of the Map task is kept on the MPP memory and, if necessary, is forwarded to a different device as part of the shuffling process.

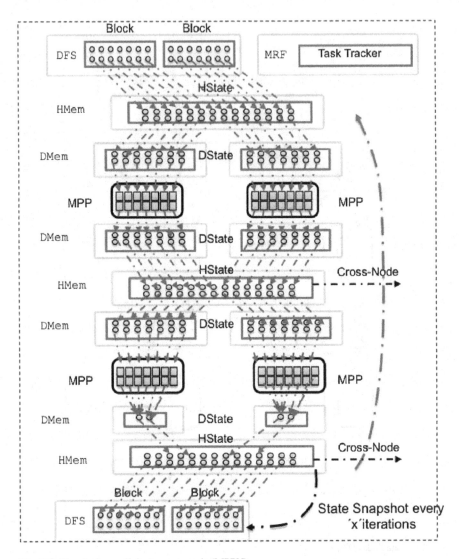

Fig. 5.7 Massively parallel processor node (MPPN)

- *Communication*: In DNs the shuffling process requires slower cross-machine communication leading to increased latency between MapReduce operators. On the contrary, the shuffling process in MPPNs is carried out through message passing between host CPU threads.
- *Iteration*: In DNs a job is terminated after the conclusion of the R step. Any additional iteration would be executed as an independent job. MPPNs provide support for iterative algorithms allowing repeatable tasks to be part of the same long-running iterative job.

5.4.3   MPP Orchestration

In general, DNs running on commodity hardware are single or dual threaded. Each CPU thread operates on a different data block, and since the results of each task are materialized to the disk, synchronization between CPU threads is not necessary. Nevertheless, the introduction of MPPs into data nodes requires the interaction of two different threading models, the classic CPU threads and the MPP threads.

As opposed to CPU threads, which are heavy, MPP threads are lighter, they have fewer registers at their disposal and will be slower, but can be launched simultaneously in groups toward the execution of the same task. Furthermore, the fact that MPP threads will run distributed across multiple devices within the same node, raises a challenge not only on the efficient coordination of thousands of these threads towards the collaborative execution of an algorithm, but also on the responsiveness and error handling of the devices running these threads.

In this section, we propose an event-driven model to orchestrate both CPU and MPP threads towards the execution of MapReduce tasks. Unlike ordinary event-driven libraries, which usually directly build upon the asynchronous operations, the method proposed in this chapter is based on the principles of Active Messages [25] and the abstraction layer provided by the Concurrency and Coordination Runtime (CCR) [7]. These abstractions are: the *Port*, the *Arbiter* and the *Dispatcher Queue*.

Figure 5.8 illustrates the three abstractions. The *Port* is an event channel in which messages are received. Posting a message to a port is a non-blocking call and the calling CPU thread continues to the next statement. The *Arbiter* decides which registered callback method should be executed to consume the message or messages. Once the method is selected, the arbiter creates the pair *(Task, Block)*, which is passed to the Dispatcher Queue associated to the port. This is an indirection that enables the creation of high-level coordination primitives. Some of the possible primitives are discussed later in this section. Each port is assigned a *Dispatcher Queue* and multiple ports can be associated with the same Dispatcher Queue. The Dispatcher Queue consists of a thread pool composed by one or more CPU threads. Available threads pick *(Task, Block)* pairs passed by the Arbiter and proceed to the execution of the task on the corresponding data block in the MPP. The MPP is stageful. It keeps a state in the memory of the device (DState) across iterations to minimize memory transfers. If necessary, it synchronizes with the state in the host memory (HState).

Next, some of the coordination primitives that can be constructed using these three abstractions are introduced:

- *Single Item Receiver*: It registers callback X to be launched when a single message of type M is received in Port A.
- *Multiple Item Receiver*: Registers callback X to be launched when n messages are received in Port A. p messages will be of type M (success) and q messages of exception type (failures), so that $p + q = n$.
- *Join Receiver*: Registers callback X to be launched when one message of type M is received in Port A and another in Port B.

Fig. 5.8 Port abstraction and
its components

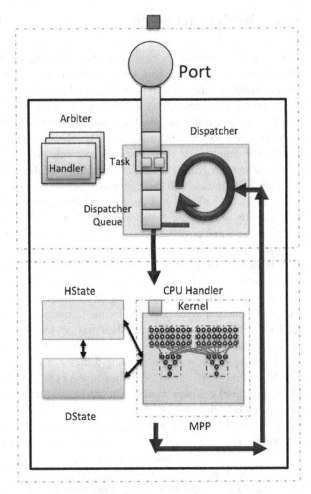

- *Choice*: Registers callback X to be launched when one message of type M is received in Port A and registers callback Y to be launched when one message of type N is received in Port A.

In the context of the MapReduce framework, these abstractions are utilized to construct a Scatter-Gather mechanism in which a master CPU thread distributes MapReduce tasks among available MPP devices, and, upon termination, these return the control and the results back to the master thread. Each MPP device will have a Port instance for every type of MapReduce task, a single Arbiter and a single Dispatcher Queue. The Arbiter will register each Port following the *Single Item Receiver* primitive with the assigned callback method that represents the MapReduce task. Requests to launch a task will contain a pointer to the data block to be manipulated and the response port in which all the responses need to be gathered. The callback method will contain the invocation of the computing kernel,

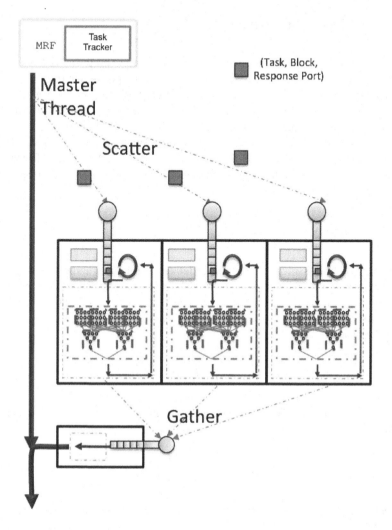

Fig. 5.9 Scatter-Gather using ports and MPPs

which spawns hundreds of MPP threads that operate the data simultaneously. The response Port follows a *Multiple Item Receiver* primitive and is registered to launch a callback method in the master thread when all the devices have answered.

This Scatter-Gather mechanism is illustrated in Fig. 5.9. One of the key benefits of this event-driven model is that not only enables coordinating multiple MPP devices towards the execution of MapReduce tasks, but also deals with the potential failure of any device, which is fundamental to conserve the robustness of the MapReduce framework.

5.5 Massively Multithreaded SVM

In order to investigate the performance of the architecture proposed in Sect. 5.4, in this section we take the SVM classifier, explore its decomposition into a MapReduce job, and launch it on a MPPN composed by multiple MPP devices. First, we provide a brief introduction to the SVM classification problem. Second, we describe the MapReduce job that solves the training phase on a single MPP device. Third, we coordinate various MPP devices using the Scatter-Gather primitive to solve a larger classification problem. Figures in this section hide the Port, Arbiter, and Dispatcher Queue components to facilitate the understanding of MapReduce task sequences.

5.5.1 Binary SVM

The binary SVM classification problem is defined as follows: Find the classification function that, given l examples $(\bar{x}_1, y_1), \ldots, (\bar{x}_l, y_l)$ with $\bar{x}_i \in R^n$ and $y_i \in \{-1, 1\}$ $\forall i$, predicts the label of new unseen samples $\bar{z}_i \in R^n$. This is achieved by solving the following regularized learning problem, where the regularization is controlled via C.

$$\min_{f \in H} C \sum_{i=1}^{l} (1 - y_i f(\bar{x}_i))_+ + \frac{1}{2} \|f\|_k^2, \tag{5.3}$$

where $(k)_+ = \max(k, 0)$. Then slack variables ξ_i are introduced to overcome the problem introduced by its non-differentiability:

$$\min_{f \in H} C \sum_{i=1}^{l} \xi_i + \frac{1}{2} \|f\|_k^2 \tag{5.4}$$

subject to: $y_i f(x_i) \geq 1 - \xi_i$ and $\xi_i \geq 0, i = 1, \ldots, l$. The dual form of this problem is given by:

$$\max_{\alpha \in R^l} \sum_{i=1}^{l} \alpha_i - \frac{1}{2} \alpha^T K \alpha \tag{5.5}$$

subject to: $\sum_{i=1}^{l} y_i \alpha_i = 0$ and $0 \leq \alpha_i \leq C, i = 1, \ldots, l$, where $K_{ij} = y_i y_j k(\bar{x}_i, \bar{x}_j)$ is a kernel function. Equation 5.5 is a quadratic programming optimization problem and its solution defines the classification function:

$$f(x) = \sum_{i=1}^{l} y_i \alpha_i k(\bar{x}, \bar{x}_i) + b, \tag{5.6}$$

where b is an unregularized bias term.

5.5.2 MapReduce Decomposition of the SVM

The binary SVM problem can be solved using the *Sequential Minimal Optimization* (SMO) algorithm [20]. SMO converts the dual form of the SVM problem into a large scale *Quadratic Programming* (QP) optimization that can be solved by choosing the smallest optimization problem at every step, which involves only two Lagrange multipliers $(\alpha_{I_{low}}, \alpha_{I_{up}})$. For two Lagrange multipliers the QP problem can be solved analytically without the need of numerical QP solvers. Next, we present SMO as an iterative sequence of MapReduce operators.

First, there are two consecutive Map operators. The first Map updates the values of the classifier function f_i based on the variation of the two Lagrange multipliers, $\Delta\alpha_{I_{low}} = \alpha'_{I_{low}} - \alpha_{I_{low}}$, $\Delta\alpha_{I_{up}} = \alpha'_{I_{up}} - \alpha_{I_{up}}$, their label values $\left(y_{I_{up}}, y_{I_{low}}\right)$ and their associated kernel evaluations:

$$f'_i = f_i + \Delta\alpha_{I_{up}} y_{I_{up}} k(\bar{x}_{I_{up}}, \bar{x}_i)$$
$$+ \Delta\alpha_{I_{low}} y_{I_{low}} k(\bar{x}_{I_{low}}, \bar{x}_i) \qquad (5.7)$$

with $i = 1 \ldots l$. The initialization values for the first Map of the iterative sequence are: $f_i = -y_i$, $\Delta\alpha_{I_{up}} = \Delta\alpha_{I_{low}} = 0$, $\alpha_{I_{low}} = \alpha_{I_{up}} = 0$, $I_{low} = I_{up} = 0$.

$$M\left[i, f_i\right] \rightarrow \left[i, f'_i\right] \qquad (5.8)$$

The second Map classifies the function values f'_i into two groups, k_{up} and k_{low}, according to these filters, in which C is the regularization parameter, $k_i \in k_{up}, k_{low}$.

$$I_0 = \{i : y_i = 1, 0 < \alpha_i < C\} \cup$$
$$\{i : y_i = -1, 0 < \alpha_i < C\} \qquad (5.9)$$
$$I_1 = \quad \{i : y_i = 1, \alpha_i = 0\} \qquad (5.10)$$
$$I_2 = \quad \{i : y_i = -1, \alpha_i = C\} \qquad (5.11)$$
$$I_3 = \quad \{i : y_i = 1, \alpha_i = C\} \qquad (5.12)$$
$$I_4 = \quad \{i : y_i = -1, \alpha_i = 0\} \qquad (5.13)$$

$$k_{up} = \{i \in I_0 \cup I_1 \cup I_2\} \qquad (5.14)$$
$$k_{low} = \{i \in I_0 \cup I_3 \cup I_4\} \qquad (5.15)$$

$$M\left[i, \alpha_i\right] \rightarrow [i, k_i] \qquad (5.16)$$

The Reduce operator takes the list of values generated by the Maps and applies a different reduction operator based on the group they belong to. For k_{up} min and arg min are used, while k_{low} requires max and arg max.

$$b_{up} = \min \{f_i : k_i = k_{up}\} \tag{5.17}$$

$$I_{up} = \arg \min_{k_i = k_{up}} f_i \tag{5.18}$$

$$b_{low} = \max \{f_i : k_i = k_{low}\} \tag{5.19}$$

$$I_{low} = \arg \max_{k_i = k_{low}} f_i \tag{5.20}$$

The indices (I_{up}, I_{low}) indicate the Lagrange multipliers that will be optimized.

$$R[k, \{f_i\}_{k_i = k}] \rightarrow [b, I] \tag{5.21}$$

The last Map uses these indices to calculate the new Lagrange multipliers:

$$\alpha'_{I_{up}} = \alpha_{I_{up}} - \frac{y_{I_{up}}(f_{I_{low}} - f_{I_{up}})}{\eta} \tag{5.22}$$

$$\alpha'_{I_{low}} = \alpha_{I_{low}} + s(\alpha_{I_{up}} - \alpha'_{I_{up}}) \tag{5.23}$$

where

$$s = y_{I_{up}} y_{I_{low}} \tag{5.24}$$

$$\eta = 2k(\bar{x}_{I_{low}}, \bar{x}_{I_{up}}) -$$
$$k(\bar{x}_{I_{low}}, \bar{x}_{I_{low}}) - k(\bar{x}_{I_{up}}, \bar{x}_{I_{up}}) \tag{5.25}$$

$$M[i, \alpha_i] \rightarrow [i, \alpha'_i] \tag{5.26}$$

Convergence is achieved when $b_{low} < b_{up} + 2\tau$, where τ is the stopping criteria.

5.5.3 Single-MPP Device SVM

As we advanced in Sect. 5.4.2, unlike single or dual core based MapReduce-like systems, MPP devices can carry out multithreaded MapReduce tasks. For the case of the single-device SVM, the data block provided by the TaskTracker represents the entire training dataset. This data block is further split into subblocks that are passed to the processors in the device. Typically, each processor can run several threads simultaneously, which enables a large number of Map or Reduce tasks being executed in parallel. Figure 5.10 schematically shows the flow of MapReduce tasks on a MPP device. Two versions of the SVM MapReduce job were

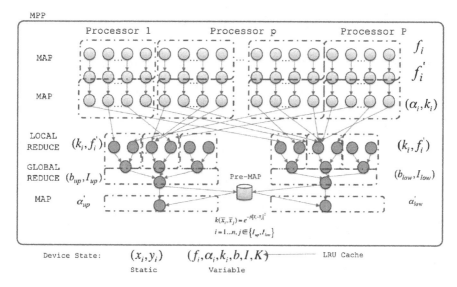

Fig. 5.10 Binary SVM decomposed into MapReduce tasks

constructed, one for each data structure type: *dense* and *sparse*. In general, sparse data structures can reduce memory utilization and data transfer times, which benefits communication within the MapReduce framework. Nevertheless, the performance of sparse algebraic operations in the MPP directly depends on the degree of sparsity of the data and the effect might be adverse, since the duplication of memory accesses caused by the additional indirection can have a negative effect on performance. The dense and sparse matrix-vector multiplications on MPPs used in this work are based on Bell et al. [2] and Vazquez et al. [24], respectively. Their impact on the SVM speed is reflected in Sect. 5.6.3.

5.5.4   Multiple-MPP Device SVM

In Sect. 5.5.3 a data block representing the entire training set was forwarded to one MPP device where MapReduce tasks would be executed iteratively until convergence, without any interaction with other devices. In this section we enhance the decomposition of MapReduce tasks to be able to break the SVM problem into multiple MPP devices. Figure 5.11 describes the interactions between four MPP devices to solve a single SVM problem. The training dataset is split into four data blocks stored in the distributed file system. The TaskTracker, that manages the master thread, forwards the corresponding block to each device. Each device performs the Map operator and a local Reduce on its local data block. The results of the reduce are gathered by the master thread, which carries out a *Global Reduce*

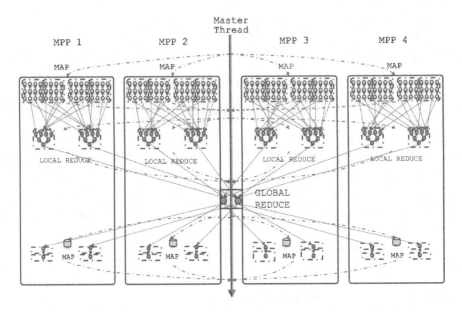

Fig. 5.11 Multiple-MPP device SVM

in order to find (b_{up}, I_{up}) and (b_{low}, I_{low}). Then these values are scattered to the devices in order to update the lagrange multipliers $\left(\alpha'_{I_{up}}, \alpha'_{I_{low}}\right)$. Finally, the master thread synchronizes, checks for convergence, and if required, proceeds to scatter the next Map task to the MPP group.

5.6 Implementation and Experimental Results

In this section we provide implementation details and performance results for the MapReduce jobs presented in Sect. 5.5, along with the incremental performance gain from one method to another. As a baseline for comparison, we take a popular SVM solver, LIBSVM [4], which is a single-threaded version of the SMO algorithm. Then, we compare LIBSVM to the SVM algorithm running on the standard Hadoop platform. Having evaluated these two popular options, we proceed to assess the performance boost obtained from the inclusion of GPUs in MapReduce cluster nodes. Throughout all the experiments the same SVM kernel functions $k(\bar{x}_i, \bar{x}_j)$, regularization parameter C, and stopping criteria τ were used.

Table 5.1 Datasets

Dataset	# Training points	# Testing points	# (Features, classes)	(C, β)
WEB	49,749	14,951	(300,2)	(64, 7.8125)
MNIST	60,000	10,000	(780,10)	(10, 0.125)
RCV1	518,571	15,564	(47,236,53)	(1, 0.1)
PROTEIN	17,766	6,621	(357,3)	(10, 0.05)
SENSIT	78,823	19,705	(100,3)	(1, 0.7)

5.6.1 Datasets

SVM training performance comparisons are carried out over five publicly available datasets, WEB [20], MNIST [18], RCV1 [19], PROTEIN [26] and SENSIT [9]. These datasets were chosen based on their computational complexity since they have hundreds of features per data sample. The sizes of these datasets and the parameters used for training are indicated in Table 5.1. The Radial Basis kernel, $k\left(\bar{x}_i, \bar{x}_j\right) = e^{-\beta \left\| \bar{x}_i - \bar{x}_j \right\|^2}$, was used for the training phase throughout all the experiments, as well as $\tau = 0.001$. Multiclass datasets, such as MNIST, RCV1, PROTEIN and SENSIT are decomposed into binary SVM problems following the One-vs-All (OVA) output code. Then, the resulting collection of binary SVMs is solved in parallel as independent MapReduce jobs.

5.6.2 Implementation and Setup

The measurements collected in the next subsection (i.e. Sect. 5.6.3) were carried out in a single machine with a dual socket Intel Xeon E5520 @ 2.26 GHz (8 cores, 16 threads) and 32 GB of RAM.

Hadoop Setup: Using this machine as a host, the SVM algorithm running on Hadoop was executed on 4 Virtual Machines (VMs), with a single core and 4 GB of RAM each. The host ran the Master Node, which contained the NameNode and the JobTracker, while the four VMs ran the DataNodes with the TaskTrackers.

Multiprocessor Setup: This machine also accommodated four GPUs. The multi-processors utilized are NVIDIA Tesla C1060 GPUs with 240 Stream Processors @ 1.3 GHz. Each GPU has 4 GB of memory and a memory bandwidth of 102 GB/s. Similar to the Hadoop case, the host machine ran the Master Node, which contained the NameNode with the JobTracker, and a MPP Node with four GPU devices. The computing kernels representing MapReduce tasks were implemented using NVIDIA CUDA.

The distribution of resources across different experiments is summarized in Table 5.2. $a + b$ represents a master threads and b device threads. We denote *SD* to the single-GPU experiment and *MD* to the multi-GPU experiment.

Table 5.2 SVM experiments

Experiment	# Host threads	# Virtual machines	# GPU devices	# GPU threads	Host mem (GB)	Device mem (GB)
LIBSVM	1	–	–	–	4	–
Hadoop	1 + 4	4	–	–	4 × 4	–
SD	1 + 1	–	1	1,024	4	4
MD	1 + 4	–	4	4,096	4 × 4	4 × 4

Table 5.3 Performance results for SVM training

Dataset (Non-zero %)		LIBSVM	Hadoop SVM	SD SVM (Dense)	SD SVM (Sparse)	MD SVM (Dense)	MD SVM (Sparse)
WEB	Time (s)	2,364.2	1,698.7	154.3	107.35	73.6	57.3
	Gain (x)	1	1.39	15.32	22.02	32.12	41.26
(3 %)	Accuracy (%)	82.69	82.69	82.69	82.69	82.69	82.69
MNIST	Time (s)	118,943.5	66,753.5	2,010.3	2,321.75	726.9	923.16
	Gain (x)	1	1.78	59.16	51.23	163.63	128.84
(19 %)	Accuracy(%)	95.76	95.76	95.76	95.76	95.76	95.76
RCV1	Time (s)	710,664	231,486	N/A	N/A	N/A	3,686
	Gain (x)	1	3.07	N/A	N/A	N/A	192.75
(0.1 %)	Accuracy(%)	94.67	94.67	N/A	N/A	N/A	94.67
PROTEIN	Time (s)	861	717.5	32.93	39.09	16.06	20.71
	Gain (x)	1	1.2	26.14	22.02	53.61	41.57
(29 %)	Accuracy(%)	70.03	70.03	70.03	70.03	70.03	70.03
SENSIT	Time (s)	8,162	4,295.78	134.670	539.32	58.29	273.96
	Gain (x)	1	1.9	60.61	15.13	140.02	29.79
(100 %)	Accuracy(%)	83.46	83.46	83.46	83.46	83.46	83.46

5.6.3 Experimental Results

In this subsection we provide the performance gain obtained by each architecture/MapReduce task flow compared to the reference implementation for all the datasets: WEB, MNIST, RCV1, PROTEIN and SENSIT. For each of the experiments we present its training time, the measured acceleration with respect to the reference implementation and the accuracy obtained from testing the calculated Support Vectors (SVs) with the test dataset. These results are collected in Table 5.3. The acceleration of the testing phase falls out of the scope this work due to its triviality.

The execution of the Map and Reduce operators, introduced in Sect. 5.5.3, on the standard Hadoop infrastructure yielded a modest performance improvement in the range of (1.20 × –3.07×) when compared to LIBSVM. Nevertheless, the results obtained from running these same operators on a SD SVM produced an order of magnitude of acceleration in the range of (15.13 × –60.61×), which is consistent with the values obtained by Catanzaro et al. [3] and Herrero-Lopez et al. [14]. Scaling out the problem to four GPUs (MD SVM) and using the GPU orchestration model presented in this paper outperformed all the previous solutions producing an overall acceleration in the range of $29.79 \times -192.75\times$. These results also show that the use of sparse data structures is beneficial for cases with high degree of sparsity (WEB and RCV1), while results adverse for the rest. The execution of the sparse MD SVM on the WEB dataset produced a $1.28\times$ gain compared to the dense MD SVM on the same dataset, while the SVM for the RCV1 dataset could not be solved on its dense SVM versions nor single device SVM form since data structures would not fit in the GPU memory. The SVM for the RCV1 dataset was solved only for the sparse MD SVM version, which produced the highest acceleration ($192.75\times$) for this set of experiments. Finally, it is necessary to point out that no accuracy loss was observed and that the same classification results were obtained on all the testing datasets across all the different systems.

5.7 Conclusions and Future Work

In this chapter, our goal was to accelerate the execution of Machine Learning algorithms running on a MapReduce cluster, while maintaining the reliability and simplicity of its infrastructure. For this purpose, we integrated massively threaded multiprocessors into the nodes of the cluster, and proposed a concurrency model that allows orchestrating host threads and thousands of multiprocessor threads spread throughout different devices so as to collaboratively solve MapReduce jobs. In order to verify the validity of this system, we decomposed the SVM algorithms into MapReduce tasks, and created a combined solution that distills the maximum degree of fine-grained parallelism. The execution of the SVM algorithm in our proposed system yielded an acceleration in the range of $29.79 \times -192.75\times$, when compared to LIBSVM and in the range of $15.68 \times -91.83\times$, when compared to the standard Hadoop implementation. To the best of our knowledge this is the shortest training time reported on these datasets for a single machine, without leaving commodity hardware nor the MapReduce paradigm. In the future, it is planned to explore the possibility of maximizing the utilization of the GPUs in the MPP Node through the execution of multiple MapReduce tasks concurrently in each device.

Acknowledgements This work was supported by the Basque Government Researcher Formation Fellowship BFI.08.80.

References

1. Apache.org: Apache mahout: scalable machine-learning and data-mining library. http://mahout.apache.org/
2. Bell, N., Garland, M.: Efficient sparse matrix-vector multiplication on CUDA. NVIDIA Technical Report NVR-2008–004, NVIDIA Corporation (2008)

3. Catanzaro, B., Sundaram, N., Keutzer, K.: Fast support vector machine training and classification on graphics processors. In: ICML'08: Proceedings of the 25th International Conference on Machine Learning, Helsinki, pp. 104–111. ACM, New York (2008). doi:http://doi.acm.org/10.1145/1390156.1390170
4. Chang, C.C., Lin, C.J.: LIBSVM: a library for support vector machines (2001). Software available at http://www.csie.ntu.edu.tw/~cjlin/libsvm
5. Chang, F., Dean, J., Ghemawat, S., Hsieh, W.C., Wallach, D.A., Burrows, M., Chandra, T., Fikes, A., Gruber, R.E.: Bigtable: a distributed storage system for structured data. In: Proceedings of the 7th USENIX Symposium on Operating Systems Design and Implementation – Volume 7, OSDI'06, Seattle, pp. 205–218. USENIX Association, Berkeley (2006)
6. Chang, E.Y., Zhu, K., Wang, H., Bai, H., Li, J., Qiu, Z., Cui, H.: Psvm: parallelizing support vector machines on distributed computers. In: NIPS (2007). Software available at http://code.google.com/p/psvm
7. Chrysanthakopoulos, G., Singh, S: An asynchronous messaging library for c#. In: Proceedings of the Workshop on Synchronization and Concurrency in Object-Oriented Languages, OOPSLA 2005, San Diego (2005)
8. Dean, J., Ghemawat, S.: Mapreduce: simplified data processing on large clusters. Commun. ACM **51**(1), 107–113 (2008). doi:http://doi.acm.org/10.1145/1327452.1327492
9. Duarte, M.F., Hu, Y.H.: Vehicle classification in distributed sensor networks. J. Parallel Distrib. Comput. **64**, 826–838 (2004)
10. Ekanayake, J., Li, H., Zhang, B., Gunarathne, T., Bae, S.H., Qiu, J., Fox, G.: Twister: a runtime for iterative mapreduce. In: Proceedings of the 19th ACM International Symposium on High Performance Distributed Computing, HPDC'10, Chicago, pp. 810–818. ACM, New York (2010)
11. Ghoting, A., Krishnamurthy, R., Pednault, E., Reinwald, B., Sindhwani, V., Tatikonda, S., Tian, Y., Vaithyanathan, S.: Systemml: declarative machine learning on mapreduce. In: Proceedings of the 2011 IEEE 27th International Conference on Data Engineering, ICDE'11, Hannover, pp. 231–242. IEEE Computer Society, Washington (2011). doi:http://dx.doi.org/10.1109/ICDE.2011.5767930.
12. Hadoop: hadoop.apache.org/core/
13. He, B., Fang, W., Luo, Q., Govindaraju, N.K., Wang, T.: Mars: a mapreduce framework on graphics processors. In: PACT'08: Proceedings of the 17th International Conference on Parallel Architectures and Compilation Techniques, Toronto, pp. 260–269. ACM, New York (2008). doi:http://doi.acm.org/10.1145/1454115.1454152
14. Herrero-Lopez, S., Williams, J.R., Sanchez, A.: Parallel multiclass classification using svms on gpus. In: GPGPU'10: Proceedings of the 3rd Workshop on General-Purpose Computation on Graphics Processing Units, Pittsburgh, pp. 2–11. ACM, New York (2010). doi:http://doi.acm.org/10.1145/1735688.1735692
15. Hillis, W.D., Steele, G.L., Jr.: Data parallel algorithms. Commun. ACM **29**(12), 1170–1183 (1986). http://doi.acm.org/10.1145/7902.7903
16. Kearns, M.: Efficient noise-tolerant learning from statistical queries. J. ACM **45**(6), 983–1006 (1998). doi: http://doi.acm.org/10.1145/293347.293351
17. Lakshman, A., Malik, P.: Cassandra: a decentralized structured storage system. SIGOPS Oper. Syst. Rev. **44**, 35–40 (2010). doi:http://doi.acm.org/10.1145/1773912.1773922
18. Lecun, Y., Bottou, L., Bengio, Y., Haffner, P.: Gradient-based learning applied to document recognition. Proc. IEEE **86**, 2278–2324 (1998)
19. Lewis, D.D., Yang, Y., Rose, T.G., Li, F.: Rcv1: a new benchmark collection for text categorization research. J. Mach. Learn. Res. **5**, 361–397 (2004)
20. Platt, J.C.: Fast training of support vector machines using sequential minimal optimization, pp. 185–208. MIT, Cambridge (1999)
21. Rafique, M.M., Rose, B., Butt, A.R., Nikolopoulos, D.S.: Cellmr: a framework for supporting mapreduce on asymmetric cell-based clusters. In: IPDPS'09: Proceedings of the 2009 IEEE International Symposium on Parallel&Distributed Processing, Rome, pp. 1–12. IEEE Computer Society, Washington (2009). doi:http://dx.doi.org/10.1109/IPDPS.2009.5161062

22. Stuart, J.A., Owens, J.D.: Multi-gpu mapreduce on gpu clusters. In: Proceedings of the 2011 IEEE International Parallel & Distributed Processing Symposium, IPDPS'11, Anchorage, pp. 1068–1079. IEEE Computer Society, Washington (2011)
23. tao Chu, C., Kim, S.K., an Lin, Y., Yu, Y., Bradski, G., Ng, A.Y., Olukotun, K.: Map-reduce for machine learning on multicore. In: Proceedings of NIPS, pp. 281–288 (2007)
24. Vazquez, F., Ortega, G., Fernandez, J., Garzon, E.: Improving the performance of the sparse matrix vector product with gpus. In: 2010 IEEE 10th International Conference on Computer and Information Technology (CIT), Bradford, pp. 1146–1151 (2010). doi:10.1109/CIT.2010.208
25. von Eicken, T., Culler, D.E., Goldstein, S.C., Schauser, K.E.: Active messages: a mechanism for integrated communication and computation. SIGARCH Comput. Archit. News **20**, 256–266 (1992)
26. Wang, J.Y.: Application of support vector machines in bioinformatics. Master's thesis, National Taiwan University, Taipei, Taiwan (2002)
27. Yoo, R.M., Romano, A., Kozyrakis, C.: Phoenix rebirth: scalable mapreduce on a large-scale shared-memory system. In: IISWC'09: Proceedings of the 2009 IEEE International Symposium on Workload Characterization (IISWC), Austin, pp. 198–207. IEEE Computer Society, Washington (2009). doi:http://dx.doi.org/10.1109/IISWC.2009.5306783

Chapter 6
Large-Scale Social Network Analysis

Mattia Lambertini, Matteo Magnani, Moreno Marzolla, Danilo Montesi, and Carmine Paolino

Abstract Social Network Analysis (SNA) is an established discipline for the study of groups of individuals with applications in several areas, like economics, information science, organizational studies and psychology. In the last fifteen years the exponential growth of online Social Network Sites (SNSs), like Facebook, QQ and Twitter has provided a new challenging application context for SNA methods. However, with respect to traditional SNA application domains these systems are characterized by very large volumes of data, and this has recently led to the development of parallel network analysis algorithms and libraries. In this chapter we provide an overview of the state of the art in the field of large scale social network analysis; in particular, we focus on parallel algorithms and libraries for the computation of network centrality metrics.

6.1 Introduction

One of the main reasons behind the enormous popularity gained by SNSs, like Facebook and Twitter, can be found in the natural tendency of humankind to create social relationships, as already condensed many centuries ago by Aristotle in his famous quote "man is by nature a political animal". By analyzing the structure of

M. Lambertini • M. Marzolla (✉) • D. Montesi
Department of Computer Science and Engineering, University of Bologna, mura A. Zamboni 7, I-40127 Bologna, Italy
e-mail: mattia@aittam.it; moreno.marzolla@unibo.it; montesi@cs.unibo.it

M. Magnani
Department of Information Technology, Uppsala University, 751 05 Uppsala, Sweden
e-mail: matteo.magnani@it.uu.se

C. Paolino
Department of Computer Science, Vrije Universiteit, Amsterdam, The Netherlands
e-mail: carmine@paolino.me

A. Gkoulalas-Divanis and A. Labbi (eds.), *Large-Scale Data Analytics*,
DOI 10.1007/978-1-4614-9242-9_6, © Springer Science+Business Media New York 2014

these networks we can acquire knowledge that is impossible to gather by focusing only on the individuals or on the single relationships that compose them.

Since the publication of Moreno's book on sociometry [42], several studies have addressed the problem of analyzing complex social structures, giving birth to the interdisciplinary research field of SNA. One of the most important results of SNA has been the definition of a set of measures that describe the role of single individuals with respect to their network of relationships. These, so-called *centrality*, measures are of great practical relevance since, e.g., they can be used to identify influential people with the potential of controlling the information flow inside communication networks. Since the pioneering work by Freeman [26], who in 1979 defined a set of geodesic centrality measures for undirected networks, the concept of centrality has proved its empirical validity several times during the years [53]. Further works, such as those by White et al. [53] and more recently by Opsahl et al. [44], extended Freeman's measures to more general network types, also with applications to recent on-line social media [51]. The main centrality metrics, i.e., degree, closeness, betweenness, and PageRank, are fundamental to increase our understanding of a network. Different metrics emphasize complementary aspects, usually related to the propagation of information [39], but also to general applications, like information searching [23] or recommendation systems. In fact, centrality measures are also used in the more general field of *Complex Network Analysis* for applications, such as studying landscape connectivity to understand the movement of organisms, analyzing proteins and gene networks, studying the propagation of diseases, and planning urban streets for optimal efficiency.

The recent proliferation of online SNSs has been of major importance for SNA. SNSs, such as blogs, multimedia hosting services and microblogging platforms, are growing incredibly fast thanks to their ability to encourage the development of *User Generated Content* through networks of users; information that is perceived as worth sharing by their users, ranging from private life to political issues, is posted online and shared within a list of connections with the potential of creating discussions. From the terrorist attack in Mumbai in 2008 to the, so-called, Twitter revolution in Iran in 2009, online SNSs have proved to be a reliable and efficient solution to communicate and spread information, becoming an interesting case study of social phenomena. Being able to compute centrality metrics on these large networks is fundamental to understand the process of information diffusion and the role of their users.

The main problem with computing centrality metrics on online social networks is the typical size of the data. As an example, Facebook has now more than 750 million active users,[1] Twitter has currently about 200 million users,[2] and LinkedIn reached more than 100 million users.[3] If we consider networks of User Generated

---

[1] http://newsroom.fb.com/content/default.aspx?NewsAreaId=22.

[2] http://www.bbc.co.uk/news/business-12889048.

[3] http://blog.linkedin.com/2011/03/22/linkedin-100-million/.

Content, Flickr has more than five billion photos[4] and in 2010 YouTube videos have been played more than 700 billion times,[5] just to list a few statistics of this kind.

The large size of current social networks (as shown in the above examples) makes their quantitative analysis challenging. From the computational point of view, SNA represents social networks as graphs on which the metrics of interest are computed. Existing sequential graph algorithms can hardly be used due to space and time constraints. These limitations motivated an increasing interest in parallel graph algorithms for SNA [35]. Initial research focused on parallel algorithms for shared memory multi-processor systems [4, 36], because graph algorithms exhibit poor locality and thus run more efficiently on shared memory architectures. However, due to the high cost of shared memory multi-processor systems, recent research is focused on algorithms for distributed memory systems [14, 27, 30, 31, 34].

This chapter focuses on the description and analysis of parallel and distributed computation of centrality measures on large networks. As stated above, centrality measures play an important role in many research areas. This, combined with the increasing size of real-world social networks and the availability of multi-core commodity computing clusters, makes parallel graph algorithms increasingly important. This chapter aims to (i) raise awareness on the existence of these powerful tools within the network analysis community; (ii) provide an overview of the most important and recent parallel graph algorithms for computation of centrality metrics; (iii) describe the main implementations of these algorithms; (iv) provide some experimental results concerning their scalability and efficiency. This work includes a general introduction on centrality measures and parallel graph algorithms, as well as up-to-date references, in order to act as a starting point for researchers seeking information on this topic.

The chapter is organized as follows. The next two sections are brief introductions to SNA and parallel computing architectures. These sections cover the main topics needed to understand the rest of the chapter and make it self-contained—readers already knowledgeable of these topics may skip them. Then we indicate the main parallel algorithms for the computation of centrality measures on large social graphs and review the main software libraries for the parallel computation of SNA centrality measures based on several criteria, so that it should be possible for professionals to match their problems with existing solutions. Finally, we present an experimental assessment of some parallel graph algorithms aimed at evaluating the scalability and efficiency of computationally demanding algorithms from the aforementioned libraries. Experimental results will help the readers to understand the potentials and limits of the current implementations, and in general will provide insights on the issues faced when dealing with graph algorithms on parallel systems. We conclude the chapter with some final remarks.

---

[4]http://blog.flickr.net/en/2010/09/19/5000000000/.

[5]http://www.youtube.com/t/press_statistics.

6.2   Social Network Analysis

A social network is a set of people interconnected by social ties, e.g., friendship
or family relationships. Although recent works on on-line SNA often use more
complex models, like multi-modal or multi-layer networks, to represent multiple
kinds of interconnected entities [20, 38], for the purpose of this chapter we will
represent a social network as a set of nodes connected by edges, where nodes
represent individuals and edges indicate their relationships.

Definition 6.1 (Social Network). A *Social Network* is a graph (V, E) where V is
a set of individuals and $E \subseteq V \times V$ is a set of social connections.

Depending on the specific SNS we may use specific graph types: undirected,
e.g., for Facebook, where friendship connections are reciprocal, directed, e.g., for
Twitter, where *following* relationships are used, and weighted, e.g., for G+, where
the strength of the connections is part of the data model. A complete treatment
of these graph types lies outside the scope of this chapter and is not necessary
to introduce centrality measures, therefore in the following we will focus only on
unweighted undirected graphs.

In this section we cover the two main aspects of social network modeling
that are needed to understand the remaining of the chapter. The first part of the
section introduces the main centrality measures: degree, closeness, betweenness
and page rank. Then, we briefly introduce the main structural models describing
the distribution of edges inside the network. This aspect is very important because
the structure of the social network may influence the efficiency of the algorithms and
real networks are not only very sparse but have also very specific internal structures.
Knowledge of these structures is thus fundamental to test the behavior of network
algorithms.

6.2.1   Centrality Measures

In this section we introduce the main centrality measures. As we have aforementioned several variations of these metrics are possible depending on the specific
kind of network. However, the objective is not to cover all possible variations but
to understand the role of the main types of centrality. At the end of the section we
provide an example with all these measures computed on a simple graph.

6.2.1.1   Degree Centrality

Degree centrality is one of the simplest centrality measures and is used to evaluate
the local importance of a vertex within the graph. This measure is defined as
the number of edges incident upon a node. Finding the individuals with high
degree centrality is important because this is a measure of their popularity inside
the network. Twitter celebrities may have hundreds of thousand followers, and

in friendship-based networks like Facebook popular users may not correspond to public figures, making degree centrality very important and yet very simple to compute. Another relevant aspect regarding degree centrality is that in real networks there is a very small proportion of users with many connections, with the consequence that user filtering based on high degrees is very effective.

Definition 6.2 (Node Degree). Given an undirected graph $G = (V, E)$, the degree $\delta(v)$ of a node $v \in V$ is the number of edges incident upon v.

Definition 6.3 ((Normalized) Degree centrality). Given a graph $G = (V, E)$, the Degree centrality $DC(v)$ of a node $v \in V$ is defined as:

$$DC(v) \stackrel{def}{=} \frac{\delta(v)}{n-1}$$

6.2.1.2 Closeness Centrality

Closeness centrality provides an average measure of the proximity between a node and all other nodes in the graph. If we assume that the longest a path the lowest the probability that an information item (meme) will be able to traverse this path, it appears how the information produced by a node with high closeness centrality will have a higher probability to reach the other nodes of the network in a short time. Differently from degree centrality, the computation of closeness centrality requires a global view of the network.

Definition 6.4 (Closeness Centrality). Given a connected graph $G = (V, E)$, let $d(u, v)$ be the distance (length of the shortest path) between $u, v \in V$. The Closeness centrality $CN(v)$ of a node $v \in V$ is defined as:

$$CN(v) \stackrel{def}{=} \frac{n-1}{\sum_{u \in V \setminus v} d(u, v)}$$

6.2.1.3 Betweenness Centrality

We have already mentioned that the information produced by popular users has a high probability of reaching many other users in the network. If these popular users have also a high value of closeness we may also expect that the process of information propagation will be fast. However, sometimes nodes that are not popular according to their degree centrality may nevertheless play an important role in propagating information: these users may act as bridges between separate sections of the network, having the potential to block the flow of information from one section to the other. These nodes are said to have a high value of betweenness centrality.

The definition of betweenness centrality is based on the concept of *pair-dependency* $\delta_{st}(v)$.

Definition 6.5 (Pair Dependency). Given a connected graph $G = (V, E)$, let σ_{st} be the number of shortest paths between $s, t \in V$ and let $\sigma_{st}(v)$ be the number of shortest paths between s and t that pass through v. Pair-dependency $\delta_{st}(v)$ is defined as the fraction of shortest paths between s and t that pass through v. Formally:

$$\delta_{st}(v) \overset{def}{=} \frac{\sigma_{st}(v)}{\sigma_{st}}$$

Betweenness centrality $BC(v)$ of a node v can be computed as the sum of the pair dependency over all pairs of vertices $s, t \in V$.

Definition 6.6 (Betweenness centrality). Given a connected graph $G = (V, E)$, the betweenness centrality $BC(v)$ of a node $v \in V$ is defined as:

$$BC(v) \overset{def}{=} \sum_{s \neq v \neq t \in V} \delta_{st}(v)$$

6.2.1.4 PageRank

A limitation of degree centrality is the fact that it considers only the direct connections of a user. However there may be users with only a few connections that are very popular. This situation would not be captured using degree centrality. A possible extension to consider higher order relationships is PageRank centrality, corresponding to the famous algorithm proposed by Larry Page, Sergey Brin, Rajeev Motwani and Terry Winograd to rank Web pages [45]. As there is a very large body of literature focusing on this measure we do not provide additional details here.

6.2.2 Summary of the Main Centrality Measures

In Fig. 6.1 we have indicated an example of the main centrality measures. In this example the node d has the highest degree centrality because it is connected to the largest number of nodes (a, c and e). At the same time it is in the middle of the network, this corresponding to a high betweenness centrality, and it is on average closer to all the other nodes, determining its high closeness centrality.

6.2.3 Network Models

Over the past fifty years several models have been proposed to explain the structure of social networks. Three models, that are often used to test network algorithms, are the *Random Network model*, the *Small-World model* and the *Scale-free model*.

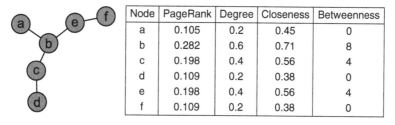

Node	PageRank	Degree	Closeness	Betweenness
a	0.105	0.2	0.45	0
b	0.282	0.6	0.71	8
c	0.198	0.4	0.56	4
d	0.109	0.2	0.38	0
e	0.198	0.4	0.56	4
f	0.109	0.2	0.38	0

Fig. 6.1 An example of the main centrality measures

The simplest model was proposed in 1959 by Erdos and Rényi, who introduced the concept of *random network* [22]. Intuitively, a random network is a graph $G = (N, E)$ in which pairs of nodes are connected with some fixed probability. A trivial algorithm to generate random graphs consists in looping through the nodes of the graph and creating an edge (v, t) with some probability p, for each pair of nodes $v, t \in E$.

Definition 6.7 (Erdös-Rényi Random network). The Erdös-Rényi model $G_{n,p}$ is an undirected graph with n nodes, where there exists an edge with probability p between any two distinct nodes.

Although very interesting and studied from a formal point of view, random networks are not representative of many real social networks: empirical evidence shows that in real networks the distance between any two nodes is lower on average than in random networks. This is often referred to as the phenomenon of the *six degrees of separation*. As a consequence, another common model used for the generation of social network datasets is the *small world model*. A small world network is a random network with the following properties:

- A short average path length.
- A high clustering coefficient.

Many real world networks expose these properties and in 1998 Watts and Strogatz proposed an algorithm to produce this kind of network [50].

The Watts-Strogatz small world network model is the best known model to represent the Small World phenomenon. However, in the last decade new important properties have been discovered concerning real networks. In 1999, Barabàsi and Albert [8] noticed that the degree distribution of many real world networks follows a power law and introduced the *scale-free model*. In simple terms, they showed that in many real-world networks there are only a few nodes with a high degree and the number of nodes with a high degree decreases exponentially. The Barabàsi-Albert model is based on the mechanism known as *preferential attachment*.[6]

In Fig. 6.2 we have represented the three aforementioned kinds of networks.

---

[6]Preferential attachment means that the probability that a new node A will be connected to an already existing node B is proportional to the number of edges that B already has.

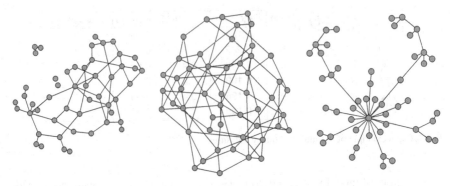

Fig. 6.2 Three graphs with 50 nodes and different structures: from left to right, a random graph (with wiring probability: 0.05), a Watts-Strogatz small world network and a Barabàsi-Albert free-scale network

6.3 Parallel Computing

Thanks to the proliferation of Web 2.0 applications and SNSs, researchers have now the opportunity to study large, real social networks. The size of those networks, which often exceeds millions of nodes and billions of edges, requires tremendous amounts of memory and processing power to be stored and analyzed.

Typically, computer programs are written for serial computation: algorithms are broken into serial streams of instructions which are executed one after another by a single processor. Parallel computing, on the other hand, uses multiple processors simultaneously to solve a single problem. To do so, the algorithms are broken into discrete parts to be executed concurrently, distributing the workload among all the computing resources, therefore using less time.

Most parallel architectures are built from commodity components instead of custom, expensive ones. In fact, the six most powerful supercomputers of the world use mass-marketed processors.[7] This makes parallel computing an efficient, cost-effective solution to large-scale computing problems, promoting its widespread adoption among both the consumer and the scientific computing worlds.

Unfortunately, there are many classes of algorithms for which an efficient parallel implementation is not known. For other classes of problems, parallel algorithms do exist but do not scale well, meaning that the algorithms are unable to make efficient use of all the resources available. Graph algorithms, which are those used in network analysis, are an example of problems which are hard to parallelize efficiently, as we will see later.

---

[7]http://www.top500.org/lists/2010/11.

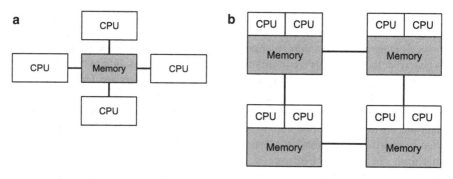

Fig. 6.3 Schematic of shared memory architectures. (**a**) UMA. (**b**) NUMA

In this section we give a general overview of parallel and distributed computing architectures and programming models, highlighting their advantages and disadvantages. We then focus our discussion on the challenges posed by parallel graph algorithms. Many books cover parallel programming and architectures in detail [17, 32]; for a recent review of the parallel computing landscape see [3].

6.3.1 Shared Memory Architectures

The last forty years have seen tremendous advances in microprocessor technology. Processor clock rates have increased from about 740 kHz (e.g., Intel 4004, circa 1971) to over 3 GHz (e.g., AMD Opteron, circa 2005). Keeping up with this exponential growth of computing power (as predicted by Moore's law [41]) has become extremely challenging as chip-making technologies are approaching fundamental physical limits [12].

Microprocessors needed a major architectural overhaul in order to improve performance. The industry found a solution to this problem in shared memory multi-processor architectures, which recently evolved into multi-core processors. In a shared memory multi-processor system there is a single, global memory space which can be accessed transparently by all processing elements. Communication and data transfer are implemented as read/write operations on the shared memory space. Shared memory multi-processors also provide suitable synchronization primitives (e.g., mutexes, memory barriers, atomic test-and-set instructions) to implement consistent updates.

Shared memory systems can be categorized by memory access times in Uniform Memory Access (UMA) and Non Uniform Memory Access (NUMA) machines. Small multi-processor systems typically have a single connection between one of the processors and the shared memory, resulting in fast and uniform memory access across all processors. Such systems are called UMA machines. Figure 6.3a shows an example of a UMA machine with four processors.

In large shared memory multi-processor systems, those having tens or hundreds of computing nodes, each processor has local and non-local (local to another processor) memory. Therefore, memory access times depend on the memory which is accessed; these systems are in fact called NUMA machines. Figure 6.3b shows an example schematic of a NUMA machine with eight processors and four memories.

Thanks to the global memory space, shared memory machines are easier to program than other parallel systems. Programmers do not have to worry about explicit data sharing and distribution among processors. Also, latencies are generally lower than in distributed memory systems (described below). However, shared memory machines are prone to hardware scalability problems—adding more processors leads to congestion of the internal connection network due to concurrent accesses to shared data structures and increasing memory write contention.

The most common way to program shared memory systems is by using threads. Threads are the smallest units of processing that can be scheduled by an operating system. In the thread programming model, each process can split into multiple concurrent executable programs. Each thread can be executed in a different processor, or in the same processor using time multiplexing, and typically shares memory with the other threads originated from the same process. Threads are commonly implemented with libraries of routines explicitly called by parallel code, such as implementations of the POSIX Threads standard, or compiler directives embedded in serial code, such as OpenMP.[8] OpenMP is a programming API for developing shared memory parallel code in Fortran, C, and C++ [43]; high level OpenMP directives can be embedded in the source code (e.g., signaling that all iterations of a "for" loop can be executed concurrently as there are no dependencies), and the compiler will automatically generate parallel code. Direct usage of threading libraries leads to greater flexibility, whereas semi-automatic generation of parallel code through OpenMP is often preferred in scientific computing thanks to its greater user-friendliness.

6.3.2 Massively Multi-threaded Computers

Massively multi-threaded computers are designed to tackle memory latency issues and context switch delays in a very different way than other parallel systems. Specifically, massively multi-threaded computers try to reduce the probability that processors become idle waiting for memory data by using a very large number of hardware threads, so that there is a chance that while some threads are blocked waiting for data, other threads can operate. Massively multi-threaded machines support very fast context switches (usually requiring a single clock cycle), dynamic load balancing, and word-level synchronization primitives. The latter is particularly important, since a large number of concurrent execution threads increases the

---

[8]http://openmp.org/.

probability of contention on shared data structures. Word-level synchronization allows the most fine control of contention.

An example of massively multi-threaded supercomputers are Cray's MTA-2 [1] and XMT [24] (also known as *Eldorado*). An MTA processor can hold the state of up to 128 instruction streams (threads) in hardware, and each stream can have up to 8 pending memory operations. Every processor executes instructions from a different non-blocked thread at each cycle, so it is fully occupied as long as there are sufficient active threads. Threads are not bound to any particular processor and the system transparently moves threads around to balance the load.

When an algorithm needs to know the global state, a synchronization must occur. On shared memory machines, synchronization is implemented in software and therefore it is an expensive operation. MTA machines solve this problem by implementing word-level synchronization primitives in hardware, drastically decreasing the cost of these operations, thus improving the system scalability.

Massively multi-threaded computers work best with algorithms that exhibit fine-grained parallelism, like most distributed graph algorithms [4]. Since the goal is to saturate the processors, the finer the level an algorithm can be parallelized, the more saturated the processors, the faster the program will run.

However, being made of custom processors on a custom architecture, MTA machines are extremely expensive and relatively slow. The latest Cray XMTs[9] includes 500 MHz Cray Threadstorm processors, which are several times slower than today's consumer multi-GHz processors. Having just 8 GB of memory per processor, Cray XMTs are also memory-constrained.

6.3.3 Distributed Memory Architectures

In contrast with the shared memory model, in the distributed memory model each processor has access to its own private memory only, and must explicitly communicate to other processors when remote data is required. A distributed memory system consists of a network of homogeneous computers, as shown in Fig. 6.4.

In a distributed memory system a process runs in a single processor, using local memory to store and retrieve data. Processes communicate using the message passing paradigm, which is typically implemented by libraries of communication routines. The de facto standard for message passing in parallel computing is Message Passing Interface (MPI) [40]. MPI is an Application Programmer Interface (API) and communication protocol standardized by the MPI Forum in 1994. It supports point-to-point and collective communication and it is language and platform-independent. Thus, many implementations of this interface exist and they are often optimized for the hardware upon which they run.

---

[9]http://www.cray.com/Products/XMT/Product/Specifications.aspx.

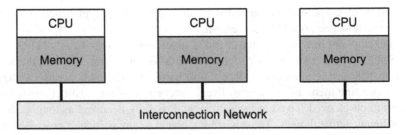

Fig. 6.4 Schematic representation of the distributed memory architecture

MPI can be combined with the thread model of computation (see Sect. 6.3.1) to exploit the characteristics of hybrid distributed/shared memory architectures, like clusters of multi-core computers.

The latest version of MPI (version 2.2) supports one-sided communication, collective extensions, dynamic process management, distributed I/O, along with many other improvements. However, only few implementations of MPI 2 exist.

Since the state of the running application is not shared across the system and every memory area is independent, distributed memory systems can be made less susceptible to interconnection overhead when scaling to hundreds, or even thousands of computing units. This is true as long as each individual node operates mostly on local data, and communication with remote nodes is not frequent.

One advantage of distributed memory architectures is that they can be built using commodity components like personal computers and consumer networks (e.g., Ethernet), whereas massive shared memory multi-processors are custom architectures. For this reason, distributed memory systems can be cost-effective when compared to other parallel architectures.

However, distributed memory systems are significantly harder to program than shared memory architectures because memory access times are non-uniform, differing several orders of magnitude from fast access to local memory, to slow network access to remote nodes. Therefore, programmers are responsible for distributing the data among all processors in a way that remote communication is reduced to a minimum. Unfortunately, as we will see shortly, this is not always possible, especially when dealing with problems that exhibit a fine-grained degree of parallelism. Graph algorithms are an example of these problems, in which most of the time is spent in fetching data from memory rather than executing CPU-bound computations.

6.3.4 Challenges in Parallel Graph Algorithms

Efficient parallel algorithms*parallel graph algorithms* have been successfully developed for many scientific applications, with particular emphasis on solving large-scale numerical problems that arise in physics and engineering [49]

(e.g., computational fluid dynamics, large N-body problems, computational chemistry, and so on). As the size of real-world networks grow beyond the capability of a single processor, a natural solution is to start looking at parallel and distributed computing as a solution to scalability problems of large scale network analysis.

Unfortunately, the development of parallel graph algorithms—the "Swiss army knife" for large scale networks analysis—faces challenges which are related to the very nature of the problem. Specifically, in [35] the authors identify the following four main challenges towards efficient parallel graph processing: (i) data-driven computations; (ii) unstructured problems; (iii) poor locality; (iv) high data access to computation ratio.

Most graph algorithms are *data-driven*, meaning that the steps performed by the algorithm are defined by the structure of the input graph, and therefore can not be predicted. This requires parallel algorithms to discover parallelization opportunities at run-time, which is less efficient than hard-coding them within the code. For example, parallel dense vector-matrix multiplication involves the same data access pattern regardless of the specific content of the vector and matrix to be multiplied; on the other hand, parallel graph algorithms (e.g., minimum spanning tree computation) may behave differently according to the input data they need to operate on.

The issue above is complicated by the fact that input data for parallel graph algorithms is usually highly *unstructured*, which makes it harder to extract parallelism by partitioning the input data. Going back to the example of vector-matrix multiplication, the input data is easily partitioned by distributing equal sized blocks (or "strides") of the input matrix to the computing nodes, such that each processor can compute a portion of the result using local data only. Partitioning a graph across all processors is difficult, because the optimal partitioning depends both on the type of algorithm executed, and also on the structure of the graph.

Graph algorithms also exhibit poor *locality of reference*. As said above, the computation is driven by the node and edge structure of the graph. In general, most graph algorithms are based on visiting the nodes in some order, starting from one or more initial locations; the neighbors of already explored nodes have larger probability to be explored next. Unfortunately, it is generally impossible to guarantee that the neighbors of a node are laid out in memory such that locality of reference is preserved. The lack of locality is particularly problematic for distributed memory architectures, where non-local data accesses are orders of magnitude slower than local (RAM) memory access. As a simple example, let us consider the exploration of a simple graph with nine nodes distributed across three different hosts. We assume that the graph forms a chain, such that there is a directed edge from node i to node $i + 1$. If we start the exploration from node 1, the partitioning shown in Fig. 6.5 causes a server to pass control to other servers for six times, while the layout shown in Fig. 6.5 causes servers to pass control two times.

Finally, graph algorithms tend to exhibit high *data access to computation ratio*. This means that a small fraction of the total execution time is spent doing computations, while most of the time is spent accessing data. Many algorithms are based on exploration of the graph rather than numerical computations, as in other types of scientific workloads. Therefore, parallel graph algorithms tend to

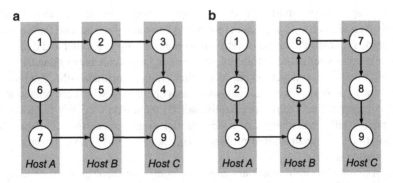

Fig. 6.5 Different partitioning strategies may lead to very different performance of the same algorithm. *Arrows* denote (directed) graph edges

be communication bound, meaning that their execution time is dominated by the cost of communication among all processors. Communication bound algorithms are difficult to parallelize efficiently, especially in parallel systems with high communication costs.

6.3.5 Addressing Locality Issues in Distributed Memory Architectures

As observed above, lack of locality and high data access to computation ratio make efficient implementation of parallel graph algorithms problematic. This problem is particularly severe for distributed memory architectures, which unfortunately are also the most appealing systems for parallel computing due to their low cost. Therefore, it is useful to spend a few words on distributed data layout for graphs.

First, we consider graphs represented using adjacency lists, which means that the graph is encoded as a set of nodes, each node having an associated set of incident edges. Alternative representations are possible, for example using adjacency matrices, where a n node graph is encoded as a $n \times n$ (usually sparse) matrix whose nonzero elements denote edges.

One possibility is to store a copy of the entire graph on each processing node (Fig. 6.6a). This has the advantage that each processor does not need to interact or pass control to other processors to explore the graph; however, synchronization with other processors might still be required by the specific algorithm being executed. Replication is particularly effective for task-parallel algorithms, that are those algorithms which can be decomposed into mostly independent tasks. For example, the computation of betweenness centrality on an n node graph using Brande's algorithm [13] requires n independent visits, each one starting from a different node. This algorithm can be efficiently implemented on distributed memory architectures by replicating the input graph and assigning to each processor the task of visiting the graph from a subset of the nodes.

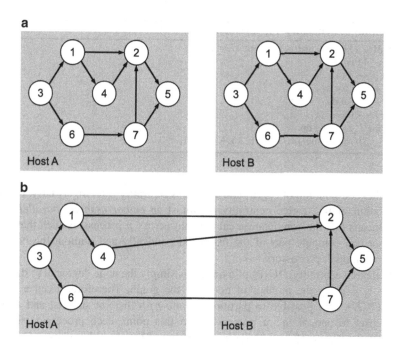

Fig. 6.6 Graph replication and partitioning example. (**a**) Replication. (**b**) Partitioning

Replication has the serious disadvantage of not making optimal use of the available memory: in fact, each computing node must have enough RAM to keep a copy of the whole graph. This is not always possible, since large real-world graphs are larger than the RAM available on any conceivable single computing node. To address this issue, it is possible to partition the input graph, such that each processing node is responsible for storing only part of the graph (Fig. 6.6). This results in the useful side effect that it is possible to handle larger graphs by simply adding more computing nodes. The downside is that communication costs might limit scalability, if the partitioning is not done accurately. Some performance results will be discussed in Sect. 6.6.

6.4 Parallel Computation of Centrality Measures

In this section we briefly describe some parallel algorithms for computation of centrality measures in social graphs. Among the metrics mentioned in Sect. 6.2, betweenness centrality is the most challenging to compute efficiently using parallel algorithms [5, 21].

Algorithm 3: Sequential Floyd-Warshall algorithm

Input: M_{ij} adjacency matrix for graph $G = (V, E)$
 for $i \leftarrow 1$ to n **do**
  for $j \leftarrow 1$ to n **do**
   $d(i, j) := M_{ij}$
 for $k \leftarrow 1$ to n **do**
  for $i \leftarrow 1$ to n **do**
   for $j \leftarrow 1$ to n **do**
    $d(i, j) \leftarrow \min(d(i, j), d(i, k) + d(k, j))$

Parallel computation of degree and closeness centrality is relatively easy. In fact, computation of the degree centrality is almost an *embarrassingly parallel* task, which means that the computation can be split across p processors such that each processor can compute part of the result independently and without the need to interact with other processors.

The degree centrality $DC(v)$ of a node v is simply the node degree $\delta(v)$ divided by $n - 1$, n being the number of nodes in the graph. Therefore, given a graph $G = (V, E)$, it is possible to partition V into n/p disjoint subsets, and assign each subset to one of the p processors. At this point, each processor trivially computes the degree of every node in its partition. This algorithm does not require any communication between processors, so it can be efficiently implemented even on distributed memory architectures with partitioned input graph. For example, host A in Fig. 6.6b can compute the degree centrality of nodes $\{1, 3, 4, 6\}$, while host B can compute the degree centrality of nodes $\{2, 5, 7\}$.

Closeness centrality $CN(v)$ involves the computation of all pairwise distances $d(u, v)$ for all nodes $u, v \in V$. Most of the existing parallel all-pair shortest path algorithms are based on either the Floyd-Warshall algorithm, or on Dijkstra's Single Source Shortest Path (SSSP) algorithm, which is executed n times, one for each node $u \in V$. In both cases, efficient parallel implementations are known [25, Chapter 3.9].

We briefly describe a parallel version of the Floyd-Warshall algorithm; the sequential version is very simple, and is described by Algorithm 3. M_{ij} is the adjacency matrix for graph G, such that M_{ij} is the length of edge (i, j), if such an edge exists, or $+\infty$ otherwise. The sequential algorithm performs a set of relaxation steps, updating the distance estimates $d(u, v)$.

A simple approach to parallelize Algorithm 3 is to partition the matrix d row wise, and assign each block of n/p contiguous rows to each of the p processors. In addition to this local data, each processor needs a copy of the k-th row $d(*, k)$ to perform its computations; at each iteration, the processor that has this row can broadcast it to all other processors. Figure 6.7 shows an example of data partitioning across four processors.

We now turn our attention to the computation of betweenness centrality. At the time of writing the best known sequential algorithm is due to Brandes [13]. If we define $\delta_s(v) = \sum_{s \neq v \neq t \in V} \delta_{st}(v)$, then the betweenness score of node $v \in V$ can be then expressed as

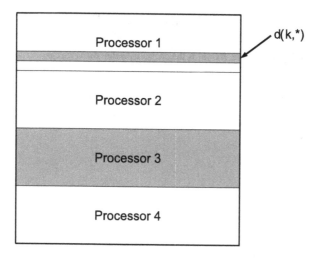

Fig. 6.7 Computing all-pair shortest paths on four processors; *shaded areas* denote the portions of matrix $d(i, j)$ which must be stored within processor 3

$$BC(v) = \sum_{s \neq v \in V} \delta_s(v)$$

Let $P_s(v)$ denote the set of predecessors of a vertex v on shortest paths from s, defined as:

$$P_s(v) = \{u \in V : (u, v) \in E, d(s, v) = d(s, u) + w(u, v)\}$$

Brandes shows that the dependencies satisfy the following recursive relation, which is the key idea of the algorithm:

$$\delta_{s*}(v) = \sum_{w, v \in P_s(w)} \frac{\sigma_{sv}}{\sigma_{sw}}(1 + \delta_{s*}(w))$$

Based on the facts above, $BC(v)$ can be computed in two steps: first, execute n SSSPs, one for each source $s \in V$, maintaining the predecessor sets $P_s(v)$; then, compute the dependencies $\delta_{s*}(v)$ for all $v \in V$. At the end we can compute the sum of all dependency values to obtain the centrality measure of v.

In 2006, Bader and Madduri [5] proposed a parallel algorithm for betweenness centrality, which exploits parallelism at two levels: the SSSP computation from each source vertex is done concurrently, and each individual SSSP is also parallelized (see Algorithm 4).

The algorithm assigns a fraction of the graph nodes to each processor, which can then initiate the SSSP computation. The stack S, list of predecessors P and the Breadth-First Search (BFS) queue Q are replicated on each computing node,

---

Algorithm 4: Bader and Madduri parallel betweenness centrality

---

Input: $G(V, E)$
Output: Array $BC[1..n]$, where $BC[v]$ is the centrality value for vertex v

  for all $v \in V$ **in parallel do**
    $BC[v] \leftarrow 0$;
  for all $s \in V$ **in parallel do**
    $S \leftarrow$ empty stack;
    $P[w] \leftarrow$ empty list, $\forall w \in V$;
    $\sigma[t] \leftarrow 0, \forall t \in V$;
    $\sigma[s] \leftarrow 1$;
    $d[t] \leftarrow -1, \forall t \in V$;
    $d[s] \leftarrow 0$;
    $Q \leftarrow$ empty queue;
    enqueue s to Q;
    while Q not empty **do**
      dequeue v from Q;
      push v to S;
      for each neighbor w of v **in parallel do**
        if $d[w] < 0$ **then**
          enqueue $w \rightarrow Q$;
          $d[w] \leftarrow d[v] + 1$;
        if $d[w] = d[v] + 1$ **then**
          $\sigma[w] \leftarrow \sigma[w] + \sigma[v]$;
          append $v \rightarrow P[w]$;
    $\delta[v] \leftarrow 0, \forall v \in V$;
    while S not empty **do**
      pop $w \leftarrow S$;
      for $v \in P[w]$ **do**
        $\delta[v] \leftarrow \delta[v] + \frac{\sigma[v]}{\sigma[w]}(1 + \delta[w])$
      if $w \neq s$ **then**
        $BC[w] \leftarrow BC[w] + \delta[w]$

---

so that every processor can compute its own partial sum of the centrality value for each vertex, and all the sums can be merged in the end using a reduction operation. Algorithm 4 is not space efficient, as it requires storing the whole SSSP tree for each source node.

The algorithm above requires fine-grained parallelism for update the shared data structures, and is therefore unsuitable for a distributed memory implementation. Edmonds, Hoefler, and Lumsdaine [21] recently proposed a new parallel space-efficient algorithm for betweenness centrality which partially addresses these issues: the new algorithm requires coarse-grained parallelism, and therefore is better suited for distributed memory architectures. Furthermore, memory requirements are somewhat reduced.

6.5 Libraries

In this section we describe some of the available software libraries and tools for graph analysis on parallel and distributed architectures—sequential network analysis packages are not considered here. The list reported here is not meant to be exhaustive. However, we tried to cover some of the most relevant parallel graph analysis packages available, both for shared memory and for distributed memory architectures. All software packages described here are distributed under free software licenses (generally, GNU General Public License or BSD-like licenses). Table 6.1 summarizes the main features of these packages.

6.5.1 Parallel Boost Graph Library

The Parallel Boost Graph Library (PBGL) [27] is a library for distributed graph computation which is part of Boost [11], a collection of open source, peer-reviewed libraries written in C++. The PBGL is based on another Boost library, the serial Boost Graph Library (BGL) [48], offering similar syntax, data structures, and algorithms. Like all the Boost libraries, it is distributed under the Boost Software License, a BSD-like permissive free software license.

The PBGL aims at being efficient and flexible at the same time by employing the generic programming paradigm. Its generic algorithms are defined in terms of collections of properties, called *concepts*. A concept defines a set of type requirements extracted from an algorithm, in order to ensure the computational feasibility and efficiency of that algorithm on user-provided types. An example is the *graph* concept, which express the fact that a graph is a set of vertices and edges with some additional identification information. This makes the PBGL flexible enough to work with parametric data structures; since those abstractions are removed by the compiler, a generic algorithm can be as fast as an hard-coded one.

In order to decouple vertex and edge properties accessors, and their representation, the PBGL implements *property maps*. Thanks to property maps, vertex and edge properties can be stored either within the graph data structure itself, or as distinct, arbitrary data structures. Property maps can be distributed in such a way that each processor only maintains the values for a subset of nodes.

The PBGL implements some important measures for social network analysis, like betweenness centrality, as well as several fundamental graph algorithms, including BFS, Depth-First Search (DFS), SSSP, Minimum Spanning Tree (MST), Connected Components (CC), and *s-t* connectivity, which determines whether there is a path from vertex *s* to vertex *t* on a given graph.

Table 6.1 Summary of parallel graph libraries

Name	Latest release	License	Lang.	Parallel API	Implemented algorithms	Input formats	Output formats
PBGL [11,27]	1.47.0 (Jul 2011)	BSD-like	C++	Distributed memory (MPI)	BFS, DFS, Dijkstra's SSSP (+ variants), MST (+ variants), CC, Strongly Connected Components (SCC), PageRank, Boman et al. graph coloring, Fruchterman Reingold force-directed layout, s-t connectivity, betweenness centrality	METIS, API	GraphViz, API
SNAP [6,37]	0.4 (Aug 2010)	GPLv2	C	Shared memory (OpenMP)	BFS, DFS, CC, SCC, diameter, vertex cover, clustering coefficient, community Kullback Liebler, community modularity, conductance, betweenness centrality, modularity betweenness, modularity greedy agglomerative, modularity spectral, seed community detection	DIMACS, GML, GraphML, METIS, SNAP, API	API
Multi-Threaded Graph Library (MTGL) [9, 47]	1.0 (Jun 2011)	BSD-like	C++	Shared memory (MTA architecture or Qthreads library)	BFS, Psearch (DFS variant), SSSP, CC, SCC, PageRank, subgraph isomorphism, random walk, s-t connectivity, betweenness centrality, community detection, connection subgraphs, find triangles, assortativity, modularity, MaxFlow	Binary, DIMACS, matrix market, memory map, API	Binary, DIMACS, matrix market, memory map, API
DisNet [19,34]	N/A (Jun 2010)	GPLv3	C++	Distributed memory (custom)	Degree centrality, betweenness centrality, closeness centrality, eccentricity	DisNet, adjacency list, Pajeck (variant), API	API
HipG [29,31]	1.5 (Apr 2011)	GPLv3	Java	Distributed memory (Ibis)	BFS, SCC, PageRank	SVC-II, Hip, API	SVC-II, Hip, API
PeGaSus [30,46]	2.0 (Sep 2010)	BSD-like	Java	Distributed memory (Hadoop)	Degree Centrality, PageRank, Random walk with restart, Radius, CC	Tab-separated	Plot
CombBLAS [14, 16]	1.1 (May 2011)	BSD-like	C++	Distributed memory (MPI-2)	Matrix Operations, Betweenness Centrality, MCL graph clustering	N/A	N/A

6.5.2 Small-world Network Analysis and Partitioning (SNAP)

SNAP [6,37] is a parallel library for graph analysis and partitioning, targeting multi-core and massively multi-threaded platforms. SNAP is implemented in C, using POSIX threads and OpenMP primitives for parallelization. SNAP is distributed under the GPLv2 software license.

SNAP pre-processes input data in order to choose the best available data structure: the default is cache-friendly adjacency arrays, but switches to dynamically resizable adjacency arrays when dynamic structural updates are required, and sorts them by vertex or edge identifier when fast deletions are necessary.

SNAP is specifically optimized for processing small-world networks by exploiting their characteristics, such as low graph diameter, sparse connectivity and skewed degree distribution. For example, SNAP represents low-degree vertex adjacencies in unsorted adjacency arrays, and high degree vertices in a structure similar to a randomized binary search tree called treap [2], for which efficient parallel algorithms for set operations exist.

On the algorithm side, SNAP fundamental graph algorithms, such as BFS, DFS, MST and CC, are designed to exploit the fine-grained thread-level parallelism offered by shared memory architectures. SNAP also supports other SNA metrics that have a linear or sub-linear computational complexity, such as average vertex degree, clustering coefficient, average shortest path length, rich-club coefficient and assortativity. These metrics are also useful as preprocessing steps for optimizing some analysis algorithms.

6.5.3 Multi-Threaded Graph Library (MTGL)

MTGL [9,10,47] is a parallel graph library designed for massively multi-threaded computers. It is written in C++ and distributed under the MTGL License, a BSD-like custom license. It is inspired by the generic programming approach of the BGL and PBGL, but rather than maximizing its flexibility, it is designed to give developers full access to the performance of MTA machines. Due to the high cost and low availability of massively multi-threaded computers, the MTGL has been extended [9] to support Qthreads [52], a library that emulates the MTA architecture on standard shared memory computers. While scalability and performance degrade when using standard shared memory architectures, Barrett et al. [9] have shown that performance issues and benefits are similar in both cases. MTGL provides a wide range of graph algorithms, including BFS, DFS, CC, SCC, betweenness centrality, community detection and many others.

6.5.4 HipG Framework

HIPG [29, 31] is a distributed framework for processing large-scale graphs. It is implemented in Java using the Ibis message-passing communication library [7], and it is distributed under the GPLv3 license. HIPG provides an high-level object-oriented interface that does not expose explicit communication. Nodes are represented by objects and can contain custom fields to represent attributes and custom methods. However, due to Java objects memory overhead, edges are stored in a single large integer array, distributed in chunks across all computing nodes; access to this data structure is managed by an abstraction layer in order to be transparent to the user. An HIPG distributed computation is defined by plain Java methods executed on graph nodes; each call to a method of a remote node (i.e., a node not stored locally on the caller processor) is automatically translated in an asynchronous message. In order to compute aggregate results, HIPG uses logical objects called *synchronizers* that can apply reduction operations. Synchronizers can also manage the computation by calling methods on nodes and setting up barriers. Synchronizers can be hierarchically organized to support sub-computations on sub-graphs, supporting divide-and-conquer graph algorithms.

6.5.5 DisNet Framework

DisNet [19, 34] is a distributed framework for graph analysis written in C++ and distributed under the GPLv3 license. Instead of relying on MPI, DisNet implements its own message passing system using standard sockets, but provides an API that abstracts away all details of parallelism. DisNet is a master-worker framework: the master coordinates the workers and combine results, and workers run a user-specified routine on vertices. Workers communicate only with the master and never between each other. DisNet does not partition the graph, so every worker has a copy of the entire network. While this ensures an high level of efficiency, especially on data structure that exhibits poor locality (like graphs, see Sect. 6.3.4), it can be a problem with networks starting with millions of nodes and billions of edges. For example, a real social network with 4,773,246 vertices and 29,393,714 edges required 10 GB of memory [34]. Aside from those potential problems, DisNet is a viable alternative to the other libraries and tools we discussed in this chapter.

6.5.6 PeGaSus System

PeGaSus [30, 46] is an open source graph mining system implemented on the top of Hadoop [28], an open source implementation of the MapReduce framework [18]. PeGaSus provides algorithms for typical graph mining tasks, such as computation of diameter, CC, PageRank and so on. The algorithms above are implemented around matrix-vector multiplications; PeGaSus provides an efficient primitive for that, called Generalized Iterated Matrix-Vector multiplication (GIM-V). The GIM-V operation on the matrix representation of a graph with n nodes and m edges requires time $O\left(\frac{n+m}{p} \log \frac{n+m}{p}\right)$ using p processors.

6.5.7 Combinatorial BLAS Library

The Combinatorial BLAS [14, 16] is a high performance distributed library for graph analysis and data mining. Like the Basic Linear Algebra Subroutines (BLAS) library [33], it defines a simple set of primitives which can be used to implement complex computations. However, the Combinatorial BLAS focuses on linear algebra primitives targeted at graph and data mining applications on distributed memory clusters. The Combinatorial BLAS implements efficient data structures and algorithms for processing distributed sparse and dense matrices, acting as a foundation to other data structures such as the sparse adjacency matrix representation of graphs. The library provides a common interface and allows users to implement a new sparse matrix storage format, without any modification to existing applications. It uses the MPI library to handle communication between computing nodes. Currently, the Combinatorial BLAS also includes an implementation of the betweenness centrality algorithm for directed, unweighted graphs, and a graph clustering algorithm.

6.6 Performance Considerations

We now present some performance results for two parallel implementations of the betweenness centrality algorithm: one for shared memory architectures and another for distributed memory architectures. We remark that the aim of this section is not to do an accurate performance analysis of the algorithms considered; instead, we want to show how the issues described in Sect. 6.3.4 can influence the scalability of parallel graph algorithms. We focus on betweenness centrality because it is a widely used metric, it is computationally demanding, and because implementations for distributed memory and shared memory architectures are readily available.

We consider two parallel implementations of the betweenness centrality algorithm. The first one is part of the PBGL [27] (included in Boost version 1.46.0) and is run on a distributed memory commodity cluster made of Intel PCs connected through a 100 Mbps fast Ethernet LAN, using MPI as the communication framework. The second implementation is part of SNAP [6] version 0.4, and is run on a single node of an IBM pSeries 575 supercomputer hosted at the CINECA

Table 6.2 Technical specifications of the machines used for the tests

	Distributed memory	Shared memory
Model	Commodity cluster	IBM pSeries 575
Nodes	61	168
CPU type	Intel Core2 Duo E7 2.93 GHz	IBM Power6 4.7 GHz
Cores per node	2	32
RAM per node	2 GB	128 GB
Network	100Mb fast Ethernet (half duplex)	Infiniband 4× DDR
Operating system	Linux 2.6.28-19	AIX 6
C/C++ compiler	GCC 4.3.3	IBM XL C/C++ 11.1.0.8
Software library	Boost 1.46.0	SNAP 0.4

supercomputing center.[10] While the IBM 575 is actually a *distributed memory* system (our installation has 158 computing nodes), a single node has 32 cores with 128 GB of shared memory, hence it can be viewed as a 32-way shared memory system. Table 6.2 describes the details of the machines used in the tests.

We study the *relative speedup* S_p of both implementations as a function of the number p of processing cores. The relative speedup is a measure of the scalability of a parallel algorithm, and is defined as follows. Let T_p be the execution time of a parallel program run on p processors; the relative speedup S_p is the ratio of the execution time on 1 processor and the execution time on p processors:

$$S_p = \frac{T_1}{T_p}$$

If the total work can be evenly partitioned across all processors and there is no communication overhead, the execution time with p processor is approximately $1/p$ times the execution time with one processor. In such (ideal) conditions, a parallel algorithm exhibits perfect (or linear) speedup $S_p \approx p$. Unfortunately, in practice we have $S_p \ll p$, due to the following reasons:

- Partitioning the input data across all processors is usually a task which must be done serially; in some cases, this task can become the bottleneck.
- If the input data is unstructured and irregular (graph data fall in this category), it may be difficult to partition it such that the workload is evenly distributed across all processors. In such cases, the slower processor may block the faster ones, introducing unwanted latencies and reducing scalability.

---

[10]http://www.cineca.it/.

- Communication and synchronization costs can become a major issue, especially with a large number of processors. At some point a parallel program may experience *negative scalability*, meaning that the execution time grows as more processors are added.

In the following sections we illustrate and discuss the results we have obtained on our test infrastructures.

6.6.1   Betweenness Centrality on Distributed Memory Architectures

We first consider the betweenness centrality algorithm on a distributed memory cluster. We use the implementation of the centrality algorithm provided by the PBGL (included in Boost 1.46.0), which uses MPI as the communication layer. The program is executed on a commodity cluster made of Intel PCs running Debian Linux, connected using Fast Ethernet. This low-end solution is poorly suited for communication-intensive high performance computing tasks; however, similar infrastructures are readily available at most institutions, so it is important to test whether they can be used for large scale network analysis.

In the following tests we used a publicly available real social network dataset with 28,250 nodes and 692,668 edges structured as a scale-free graph [15]. The size of the graph is motivated by the need to get reasonable computation times: computing the centrality values of all vertices on a single CPU core requires about 5 h on our infrastructure. The input graph was fully replicated on each node.

We also used an increasing number p of computing nodes, from $p = 1$ to $p = 8$. Larger values of p do not yield any significant advantage on the graph above. We run a single process on each machine (although all processors are dual core), since we do not want to mix local (inter-node) communications with remote (intra-node) ones. The execution times T_p with p processing nodes have been computed by averaging the results obtained over multiple independent experiments. Of course, all tests have been executed when all nodes were idle.

Figure 6.8 shows the speedup S_p and average execution times T_p with p processors. The algorithm achieves optimal (linear) speedup with the replicated input graph. This result is quite remarkable, considering that the test cluster has a poor network connection. The reason is that the parallel centrality algorithm has a low communication to computation ratio: each processor is assigned the computation of an SSSP, which can be computed using local data only. Inter-node communications happen only at the end of each SSSP, when each node updates the common data structure of centrality estimates. At the end of the algorithm, this data structure contains the exact centrality values for each node.

To achieve good scalability on the cluster, it is essential to reduce communications as much as possible. If the input graph is distributed across all processors, the situation changes dramatically. Figure 6.9 shows the execution time of the

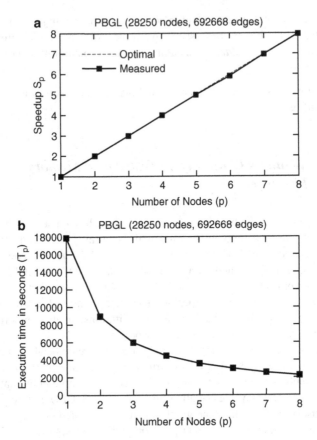

Fig. 6.8 Speedup and execution time of the betweenness centrality algorithm in Boost, executed on the Intel cluster. The input graph (28,250 nodes, 692,668 edges) is replicated across the computing nodes. (**a**) Speedup. (**b**) Execution time

betweenness centrality algorithm on the same cluster as above, using a distributed storage model for the input graph. This means that each host only stores a subset of the graph; therefore, each SSSP computation must pass control to different processors, as the graph node being visited at each step may reside on a remote host. In this situation we observe negative scalability, since the execution time grows as more processors are added. Code profiling confirms that the algorithm is communication bound, which means that a significant fraction (about 90 % in our case) of the execution time is spent waiting for data to be sent or received through the slow network.

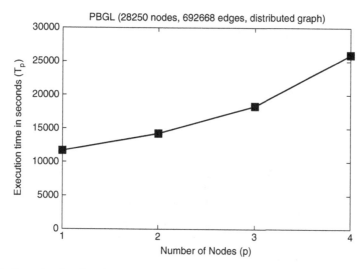

Fig. 6.9 Execution time for a Small-World graph with 28,250 nodes and 692,668 edges. The input graph is distributed across the computing nodes

6.6.2 Betweenness Centrality on Shared Memory Architectures

We now study the scalability of a different implementation of the betweenness centrality algorithm. Specifically, we consider the implementation provided by SNAP version 0.4 for shared memory architectures. The algorithm has been applied to the same graph (28,250 nodes and 692,668 edges), and has been executed on a single node of the IBM pSeries 575 supercomputer. The node has 32 processor cores sharing a common memory, therefore it can be seen as a shared memory system.

Figure 6.10 shows the speedup and total execution time with p processor cores, $p \in \{1, 2, 4, 8, 16\}$. The scalability is very limited: with $p = 16$ cores, the algorithm requires about 47 % the time required with a single core. In Fig. 6.11 we show the performance results of the betweenness centrality algorithm of SNAP on a larger graph with 224288 nodes and 3 million edges.

While communication on a shared memory multi-processor is much more efficient than in commodity distributed memory clusters, memory access is still a bottleneck for data-intensive applications, since the memory bandwidth can quickly become inadequate to feed all processors.

6.6.3 Shared Memory vs. Distributed Memory

We conclude this section by reporting the results of a direct comparison between the PBGL and SNAP, both running on the commodity cluster, using the graph with

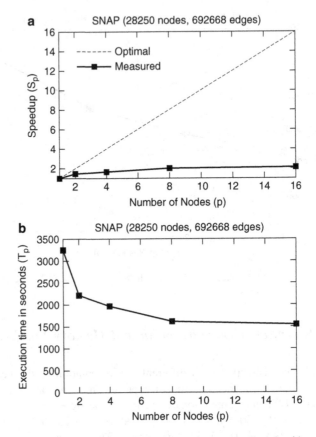

Fig. 6.10 Speedup and execution time of the betweenness centrality algorithm in SNAP on the IBM p575, for a Small-World graph with 28,250 nodes and 692,668 edges. (**a**) Speedup. (**b**) Execution time

28,250 nodes and 692,668 edges. The distributed memory version is executed on p processors, from $p = 1$ to $p = 8$, using full replication of the input graph on all hosts. SNAP has been executed on a single host of the cluster using both processor cores, requiring about $1,400\,s$ to process the whole graph.

Figure 6.12 shows the execution times of the two programs. It is interesting to observe that SNAP, using only two cores of a single CPU, is faster than PBGL running on $p = 8$ nodes for this graph. Scalability is an important metric, but should not be considered alone: a less scalable algorithm may be faster in practice than a perfectly scalable one, as this test demonstrates.

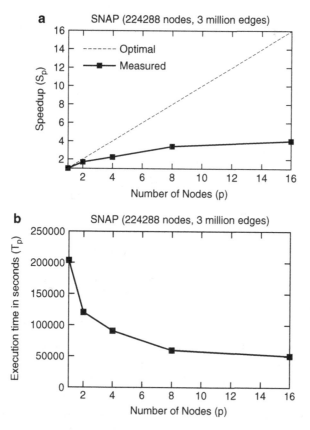

Fig. 6.11 Speedup and execution time of the betweenness centrality algorithm in SNAP on the IBM p575, for a larger Small-World graph with 224,288 nodes and 3 million edges. (**a**) Speedup. (**b**) Execution time

6.7 Concluding Remarks

In this chapter we considered the problem of computing centrality measures in large social networks using parallel and distributed algorithms. We first introduced the main centrality measures used in SNA. Then, we gave an overview of the main features and limitations of current parallel and distributed architectures, including distributed memory, shared memory and massively multi-threaded machines. We then briefly described some of the existing parallel algorithms for computing useful centrality measures. After that, we presented a set of software packages that implement such algorithms. Finally, we gave some insights on the performance of the betweenness centrality algorithms provided by the PBGL (on distributed memory architectures), and by SNAP (on shared memory architectures).

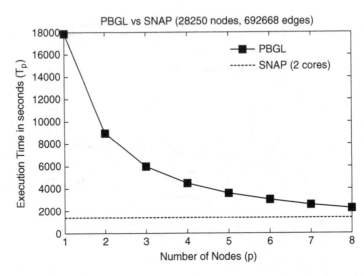

Fig. 6.12 Execution time of the betweenness centrality algorithm provided by the PBGL and by SNAP. Both algorithms have been run on the commodity cluster; SNAP has been executed on a single server using both CPU cores

Parallel graph algorithms are a hot research topic, which is still waiting to make a real breakthrough. Graph algorithms exhibit a number of properties which make them very hard to parallelize efficiently on current high performance computing architectures: the large size of social graphs makes it very difficult to store them in RAM, and distributed storage of graph data across multiple computing nodes raises other issues, related to the long access times to remote graph data. Graph algorithms also exhibit a large communication to computation ratio, making them poor candidates for parallel implementation.

Despite the issues above, more efficient parallel architectures, combined with state-of-the-art algorithms, may open the possibility of managing large social network graphs, enabling scientists to get a better understanding of the social phenomena which, directly or indirectly, influence our lives.

Acknowledgements This work has been partially funded by PRIN project *"Relazioni sociali e identità in rete: vissuti e narrazioni degli italiani nei siti di social network"* and by FIRB project *"Information monitoring, propagation analysis and community detection in Social Network Sites"*. This work was done while M. Magnani and C. Paolino were with the Deptartment of Computer Science, University of Bologna.

The authors thank the CINECA supercomputing center for providing access to the IBM pSeries 575 used for part of the tests described in Sect. 6.6.

References

1. Anderson, W., Briggs, P., Hellberg, C.S., Hess, D.W., Khokhlov, A., Lanzagorta, M., Rosenberg, R.: Early experience with scientific programs on the cray MTA-2. In: Proceedings of 2003 ACM/IEEE Conference on Supercomputing, SC'03, Phoenix, p. 46. ACM, New York, (2003). doi:10.1145/1048935.1050196

2. Aragon, C.R., GSeidel, R.: Randomized search trees. In: Annual IEEE Symposium on Foundations of Computer Science, Research Triangle Park. IEEE Computer Society, Los Alamitos, pp 540–545 (1989). doi:10.1109/SFCS.1989.63531

3. Asanovic, K., Bodik, R., Demmel, J., Keaveny, T., Keutzer, K., Kubiatowicz, J., Morgan, N., Patterson, D., Sen, K., Wawrzynek, J., Wessel, D., Yelick, K.: A view of the parallel computing landscape. Commun ACM 52, 56–67 (2009)

4. Bader, D.A., Madduri, K.: Designing multithreaded algorithms for breadth-first search and st-connectivity on the Cray MTA-2. In: Proceedings of International Conference on Parallel Processing, Columbus. IEEE Computer Society, Los Alamitos, pp 523–530 (2006). doi:10.1109/ICPP.2006.34

5. Bader, D.A., Madduri, K.: Parallel algorithms for evaluating centrality indices in real-world networks. In: Proceedings of 2006 International Conference on Parallel Processing, ICPP'06, Columbus, pp. 539–550. IEEE Computer Society, Washington, DC (2006). doi:10.1109/ICPP.2006.57

6. Bader, D.A., Madduri, K.: SNAP, small-world network analysis and partitioning: an open-source parallel graph framework for the exploration of large-scale networks. In: Proceedings of International Symposium on Parallel and Distributed Processing, IPDPS, Miami, pp. 1–12 (2008). doi:10.1109/IPDPS.2008.4536261

7. Bal, H.E., Maassen, J., van Nieuwpoort, R.V., Drost, N., Kemp, R., Palmer, N., Wrzesinska, G., Kielmann, T., Seinstra, F., Jacobs, C.: Real-world distributed computing with Ibis. Computer 43, 54–62 (2010). doi:10.1109/MC.2010.184

8. Barabasi, A.L., Albert, R.: Emergence of scaling in random networks. Science 286(5439), 11 (1999)

9. Barrett, B.W., Berry, J.W., Murphy, R.C., Wheeler, K.B.: Implementing a portable multi-threaded graph library: the MTGL on Qthreads. In: IEEE International Symposium on Parallel & Distributed Processing, IPDPS, Rome, pp. 1–8 (2009). doi:10.1109/IPDPS.2009.5161102

10. Berry, J.W., Hendrickson, B., Kahan, S., Konecny, P.: Graph software development and performance on the MTA-2 and Eldorado. In: 48th Cray Users Group Meeting, Lugano (2006)

11. Boost: Boost C++ Libraries. Available at http://www.boost.org/ (2011)

12. Borkar, S.: Design challenges of technology scaling. IEEE Micro 19(4), 23–29 (1999)

13. Brandes, U.: A faster algorithm for betweenness centrality. J. Math. Sociol. 25, 163–177 (2001)

14. Buluç, A., Gilbert, J.R.: The combinatorial BLAS: design, implementation, and applications. Int. J. High Perform. Comput. Appl. 25, 496–509 (2011). doi:10.1177/1094342011403516

15. Celli, F., Di Lascio, F., Magnani, M., Pacelli, B., Rossi, L.: Social network data and practices: the case of friendfeed. In: Chai, S.K., Salerno, J., Mabry, P. (eds.) Advances in Social Computing. LNCS, vol. 6007, pp 346–353. Springer, Berlin/Heidelberg (2010). doi:10.1007/978-3-642-12079-4_43

16. Combinatorial BLAS: Combinatorial BLAS Library (MPI reference implementation). Version 1.1, Available at http://gauss.cs.ucsb.edu/~aydin/CombBLAS/html/index.html (2011)

17. Culler, D., Singh, K.P., Gupta, A.: Parallel Computer Architecture – A Hardware/Software Approach. Morgan Kaufmann, San Francisco (1998)

18. Dean, J., Ghemawat, S.: Mapreduce: a flexible data processing tool. Commun. ACM 53(1), 72–77 (2010). doi:10.1145/1629175.1629198

19. DisNet: DisNet, A Framework for Distributed Graph Computation. Available at http://nd.edu/~dial/software.html (2011)

20. Du, N., Wang, H., Faloutsos, C.: Analysis of large multi-modal social networks: patterns and a generator. In: Proceedings of the 2010 European conference on Machine Learning and Knowledge Discovery in Databases: Part I, ECML PKDD'10, Barcelona, pp. 393–408. Springer, Berlin/Heidelberg, (2010). http://portal.acm.org/citation.cfm?id=1888258.1888291

21. Edmonds, N., Hoefler, T., Lumsdaine, A.: A space-efficient parallel algorithm for computing betweenness centrality in distributed memory. In: Proceedings of International Conference on High Performance Computing (HiPC), Dona Paula, pp. 1–10 (IEEE, 2010). doi:10.1109/HIPC.2010.5713180

22. Erdős, P., Rényi, A.: On random graphs I. Publ Math Debrecen 6, 290–297, 156 (1959)

23. Evans, B.M., Chi, E.H.: Towards a model of understanding social search. In: Proceedings of the 2008 ACM Conference on Computer Supported Cooperative Work, CSCW '08, San Diego. ACM, New York, pp. 485–494 (2008). doi:10.1145/1460563.1460641

24. Feo, J., Harper, D., Kahan, S., Konecny, P.: Eldorado. In: Proceedings of 2nd Conference on Computing Frontiers, CF '05, Ischia. ACM, New York, pp. 28–34 (2005)

25. Foster, I.: Designing and Building Parallel Programs: Concepts and Tools for Parallel Software Engineering. Addison-Wesley Longman, Boston (1995)

26. Freeman, L.C.: Centrality in social networks: a conceptual clarification. Soc. Netw. 1(3), 215–239 (1978–1979)

27. Gregor, D., Lumsdaine, A.: The Parallel BGL: A generic library for distributed graph computations. In: Parallel Object-Oriented Scientific Computing, POOSC, Glasgow (2005)

28. Hadoop.: Apache hadoop. Available at http://hadoop.apache.org/ (2011)

29. HipG.: HipG: High-level distributed processing of large-scale graphs. Available at http://www.cs.vu.nl/~ekr/hipg/ (2011)

30. Kang, U., Tsourakakis, C.E., Faloutsos, C.: PEGASUS: mining peta-scale graphs. Knowl. Inf. Syst. 27(2), 303–325 (2011). doi:10.1007/s10115-010-0305-0

31. Krepska, E., Kielmann, T., Fokkink, W., Bal, H.: A high-level framework for distributed processing of large-scale graphs. In: Proceedings of the 12th International Conference on Distributed Computing and Networking, ICDCN'11, Bangalore, pp. 155–166. Springer, Berlin/Heidelberg (2011)

32. Kumar, V., Gupta, A.G.A., Karpis, G.: Introduction to Parallel Computing, 2nd edn. Addison Wesley, Harlow (2003)

33. Lawson, C.L., Hanson, R.J., Kincaid, D.R., Krogh, F.T.: Basic linear algebra subprograms for fortran usage. ACM Trans Math Softw 5, 308–323 (1979). doi:10.1145/355841.355847

34. Lichtenwalter, R.N., Chawla, N.V.: DisNet: A framework for distributed graph computation. In: Proceedings 2011 International Conference on Social Networks Analysis and Mining (ASONAM), Kaohsiung (2011, to appear)

35. Lumsdaine, A., Gregor, D., Hendrickson, B., Berry, J.W.: Challenges in parallel graph processing. Parallel Process. Lett. 17(1), 5–20 (2007)

36. Madduri, K., Bader, D.A.: Compact graph representations and parallel connectivity algorithms for massive dynamic network analysis. In: Proceedings of International Parallel and Distributed Processing Symposium, IPDPS, Rome. IEEE Computer Society, Los Alamitos, pp. 1–11 (2009)

37. Madduri, K., Bader, D.A.: Small-world Network Analysis and Partitioning–Version 0.4. Available at http://snap-graph.sourceforge.net/ (2010)

38. Magnani, M., Rossi, L.: The ml-model for multi layer network analysis. In: IEEE International Conference on Advances in Social Network Analysis and Mining, Kaohsiung (2011)

39. Magnani, M., Rossi, L., Montesi, D.: Information propagation analysis in a social network site. In: 2010 International Conference on Advances in Social Networks Analysis and Mining, Odense, pp. 296–300. IEEE Computer Society, Los Alamitos (2010)

40. Message Passing Interface Forum MPI: A Message-Passing Interface Standard–Version 2.2. Available at http://www.mpi-forum.org/docs/ (2009)

41. Moore, G.E.: Cramming more components onto integrated circuits. Proc. IEEE 86(1), 82 (1998). doi:10.1109/JPROC.1998.658762

42. Moreno, J.L., Jennings, H.H.: Who Shall Survive? : A New Approach to the Problem of Human Interrelations. Nervous and Mental Disease Publishing Co., Washington, D.C. (1934)
43. OpenMP Architecture Review Board: OpenMP Application Program Interface–Version 3.1. Available at http://openmp.org/wp/ (2011)
44. Opsahl, T., Agneessens, F., Skvoretz, J.: Node centrality in weighted networks: Generalizing degree and shortest paths. Soc. Netw. **32**(3), 245–251 (2010)
45. Page, L., Brin, S., Motwani, R., Winograd, T.: The pagerank citation ranking: bringing order to the web. Technical report, Stanford Digital Library Technologies Project (1998)
46. Pegasus: Project Pegasus. Available at http://www.cs.cmu.edu/~pegasus/ (2011)
47. Sandia National Laboratories: Multi-Threaded Graph Library–Version 1.0. Available at https://software.sandia.gov/trac/mtgl (2011)
48. Siek, J., Lee, L.Q., Lumsdaine, A.: The Boost Graph Library: User Guide and Reference Manual. Addison-Wesley, Boston (2002)
49. Trobec, R., Vajteršic, M., Zinterhof, P. (eds.): Parallel Computing: Numerics, Applications, and Trends. Springer, Dordrecht/New York (2009). doi:10.1007/978-1-84882-409-6_1
50. Watts, D.J., Strogatz, S.H.: Collective dynamics of "small-world" networks. Nature **393**(6684), 440–442 (1998)
51. Weng, J., Lim, E.P., Jiang, J., He, Q.: Twitterrank: finding topic-sensitive influential twitterers. In: Proceedings of Third ACM International Conference on Web Search and Data Mining, WSDM '10, New York, pp. 261–270. ACM, New York (2010). doi:10.1145/1718487.1718520
52. Wheeler, K.B., Murphy, R.C., Thain, D.: Qthreads: an api for programming with millions of lightweight threads. In: 22nd IEEE International Symposium on Parallel and Distributed Processing, IPDPS. IEEE, Miami, pp. 1–8 (2008). doi:10.1109/IPDPS.2008.4536359
53. White, D., Borgatti, S.: Betweenness centrality measures for directed graphs. Soc. Netw. **16**(4), 335–346 (1994). doi:10.1016/0378-8733(94)90015-9

Chapter 7
Visual Analysis and Knowledge Discovery for Text

Christin Seifert, Vedran Sabol, Wolfgang Kienreich, Elisabeth Lex, and Michael Granitzer

Abstract Providing means for effectively accessing and exploring large textual data sets is a problem attracting the attention of text mining and information visualization experts alike. The rapid growth of the data volume and heterogeneity, as well as the richness of metadata and the dynamic nature of text repositories, add to the complexity of the task. This chapter provides an overview of data visualization methods for gaining insight into large, heterogeneous, dynamic textual data sets. We argue that visual analysis, in combination with automatic knowledge discovery methods, provides several advantages. Besides introducing human knowledge and visual pattern recognition into the analytical process, it provides the possibility to improve the performance of automatic methods through user feedback.

7.1 Introduction

The already huge amount of electronically available information is growing further at an astonishing rate: an IDC study [12] estimates that by 2006 the amount of digital information exceeded 161 Exabyte, while an updated forecast [13] estimates that by 2012 the amount of information will double every 18 months. While retrieval tools excel at finding a single, or a few relevant pieces of information, scalable analysis techniques, considering large data sets in their entirety, are required when a holistic view is needed.

Knowledge discovery (KD) is the process of automatically processing large amounts of data to identify patterns and extract useful new knowledge [9].

C. Seifert (✉) • M. Granitzer
University of Passau, 94030 Passau, Germany
e-mail: christin.seifert@uni-passau.de; michael.granitzer@uni-passau.de

V. Sabol • W. Kienreich • E. Lex
Know-Center Graz, Inffeldgasse 13/6, A-8010 Graz, Austria
e-mail: vsabol@know-center.at; wkien@know-center.at; elex@know-center.at

A. Gkoulalas-Divanis and A. Labbi (eds.), *Large-Scale Data Analytics*,
DOI 10.1007/978-1-4614-9242-9_7, © Springer Science+Business Media New York 2014

KD was traditionally applied on structured information in databases; however, as information is increasingly present in unstructured or weakly structured form, such as text, adequate techniques were developed. The shift from large, static, homogeneous data sets to huge, dynamic, heterogeneous repositories necessitates approaches involving both automatic processing and human intervention. Automatic methods put the burden on machines, but despite algorithmic advancements and hardware speed-ups, for certain tasks, such as pattern recognition, human capabilities remain unchallenged.

Information visualization techniques rely on the powerful human visual system, which can recognize patterns, identify correlations and understand complex relationships at once, even in large amounts of data. Visualization is an effective enabler for exploratory analysis [52], making it a powerful tool for gaining insight into unexplored data sets.

Visual Analytics is an interdisciplinary field based on information visualization, knowledge discovery and cognitive and perceptual sciences, which deals with designing and applying interactive visual user interfaces to facilitate analytical reasoning [50]. It strives for tight integration between computers, which perform automatic analysis, and humans, which steer the process through interaction and feedback. Combining the advantages of visual methods with automatic processing provides effective means for revealing patterns and trends, and unveiling hidden knowledge present in complex data [23,48]. Analytical reasoning is supported based on the discovered patterns, where users can pose and test a hypothesis, provide assessments, derive conclusions and communicate the newly acquired knowledge.

Especially for large text repositories, Visual Analytics is a promising approach. Turning textual information into visual representations allows to access large document repositories using the human pattern recognition abilities. Providing Visual Analytics environments for text requires text mining and text analysis algorithms in order to extract information and metadata. Further, appropriate representations have to be devised in order to visualize the aspects interesting to the task at hand.

In this chapter, we provide an overview on visual analysis techniques for textual data sets, outline underlying processing elements and possible application scenarios. First, the general processing pipeline for Visual Analytics in text repositories is outlined in Sect. 7.2, followed by a detailed description of all the necessary steps. Second, Sect. 7.3 describes how visual representations can be used on the extracted information. Different visualizations are represented depending on the data aspect to be visualized. Section 7.3.1 describes topical overviews, Sect. 7.3.2 representations for multi-dimensional data, Sect. 7.3.3 spatio-temporal visualizations, and Sect. 7.3.4 visualization of arbitrary relations. The concept of user feedback integration, as well as examples, are covered in Sect. 7.3.5. The concept of combining multiple visualizations into one interface, as multiple coordinated views, is explained in Sect. 7.3.6. Third, Sect. 7.4 describes three applications: media analysis (Sect. 7.4.1), visual access to encyclopedias (Sect. 7.4.2) and patent analysis (Sect. 7.4.3). Finally, Sect. 7.5 concludes the chapter and provides an outlook of future developments in the field of Visual Analytics focusing on text data.

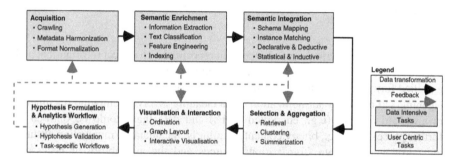

Fig. 7.1 The processing pipeline for visual analysis of text combines data-intensive tasks (*top*) and user-centric tasks (*bottom*). *Solid black lines* indicate data flows while *dashed red lines* indicate user feedback to adapt automatic processes

7.2 Processing Pipeline for Visual Analysis of Text

Visual analytics combines information visualization techniques with knowledge discovery methods in an iterative fashion. Starting from a given data set, mining techniques identify interesting, non-trivial patterns, which may provide insights on the data set. In a discovery task, where the aim is to identify new, potentially useful insights, a priori assumptions underlying the mining techniques may not be fulfilled. By visualizing the extracted patterns, humans are empowered to incorporate their background knowledge into the automatic processes through identifying wrong assumptions and erroneously identified patterns. Information visualization serves as the communication channel between the user and the mining algorithm, allowing domain experts to control the data-mining process, to rule-out wrong mining results or to focus on particularly interesting sub-samples of the data set.

Providing Visual Analytics environments for text requires text mining and text analysis algorithms, in order to extract meaningful patterns subject to visualization. The data set usually consists of a set of documents and additional metadata. Acquiring and processing these metadata is usually composed of consecutive steps resembling the traditional knowledge discovery chain. Visual analytics aims to provide users with intelligent interfaces controlling parts of these steps in order to gain new insights. Understanding individual steps is necessary to derive suitable visualizations and interactions for each step. Hence, we outline important details on the process in the following and afterwards derive potential visualizations and interactions for conducting visual analytics tasks in large text repositories. Figure 7.1 depicts an overview of the outlined process.

7.2.1 Acquisition

Acquisition includes crawling and accessing repositories to collect information, and document pre-processing, such as harmonization of metadata and conversion into a unified format. An often underestimated effort within the acquisition step is data cleaning and metadata harmonization. Data cleaning involves removing documents in awkward formats (e.g., encrypted PDFs) and data that should be omitted in further processing (e.g., binary content). Similarly, metadata harmonization ensures the correctness of data from various sources and the availability of necessary information for later semantic integration [2].

7.2.2 Semantic Enrichment

Semantic enrichment extracts domain-specific semantics from single documents and enriches each document with external knowledge. Usually, the process starts with annotating the document content with linguistic properties like part-of-speech or punctuation, or external knowledge like thesauri concepts. Annotated text then serves as basis for either extracting explicit metadata on document level, e.g., the author of a document, or to generate features representing the document in subsequent analysis steps. Annotations, metadata and features may serve as input for creating index structures to enable fast, content-based access to documents. Figure 7.2 provides an overview on these data transformations happening during semantic enrichment. In the following, we outline the most important techniques in detail.

7.2.2.1 Information Extraction

Information Extraction (IE) deals with extracting structured information from unstructured, or weakly structured, text documents using natural language processing methods [20]. IE decomposes text into building blocks, generates annotations and extracts metadata, typically employing the following methods:

(i) Tokenization, sentence extraction and part-of-speech (POS) tagging (i.e. recognizing nouns and verbs)
(ii) Named entity recognition identifies entities such as persons, organizations, locations, numbers (e.g.,time, money amounts). Co-reference detection identifies various spellings or notations of a single named entity.
(iii) Word sense disambiguation identifies the correct sense of a word depending on its context.
(iv) Relationship discovery identifies relations, links and correlations.

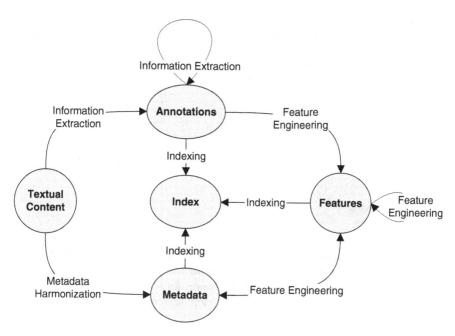

Fig. 7.2 Semantic enrichment steps starting at single artifacts, e.g., documents, news articles, patents (*left*), and resulting in enriched representations in an index (*center*)

7.2.2.2   Feature Engineering and Vectorization

Feature engineering and vectorization uses IE results and document metadata to identify, weight (e.g., TF-IDF), transform (e.g., stemming) and select (e.g., stop-word filtering) features describing text documents. Features are represented as feature vectors used by algorithms to compare documents and compute document similarities. Multiple feature spaces (also referred to as *feature name spaces*) group features of similar characteristic to describe different, potentially orthogonal aspects of a document. For example, one feature space can capture all nouns, while a second feature space can capture all extracted and pre-defined locations and a third all persons. By separating feature spaces, subsequent algorithms can take care of different feature distributions and consider different importance among feature spaces depending on the analytical task at hand, as e.g., in [32]. Besides data integration and cleaning, good feature engineering becomes the second most important step in every Visual Analytics workflow and hence subject for being steered in the analytical process.

Table 7.1 Levels of and techniques used for semantic integration

		Level	
		Schema	Instance
Technique	Declarative-deductive	Shared vocabulary, reasoning-based integration	Shared identifiers (e.g. URIs), rule-based transformation, declarative languages and identifier schemas
	Statistical-inductive	Similarity based on structure, linguistic or data type	Similarity estimates based on entity properties (e.g. clustering, near-duplicate search)

7.2.2.3 Indexing

Indexing develops efficient index structures in order to search for documents containing particular features, or sequences/annotations of features themselves. Inverted indices, representing for each feature the list of occurrences in documents, are among the most often used indexing structures for text. They exploit power law distributions of features in order to allow efficient search and retrieval in text based repositories [34].

7.2.2.4 Text Classification

Text classification employs supervised machine learning methods to organize documents into a predefined set of potentially structured categories [40]. Classification can be seen as injecting structured knowledge via a statistical, inductive process. Examples are assigning documents to topical categories, estimating genre information or determining the sentiment of text passages.

7.2.3 Semantic Integration

Semantic integration aims at integrating information from different, potentially decentralized information sources based on information provided by each source and by previous semantic enrichment processes. With the advancement of decentralized information systems, like the Web, semantic integration becomes a more and more important topic. Semantic integration, also known as *data fusion* in the database community [2], or *ontology mediation* in the Semantic Web community [3], takes place on two levels: the *schema-level* and the *instance-level*. Orthogonal to the two levels, techniques used for integration can be distinguished into *declarative-deductive* and *statistical-inductive* techniques (see Table 7.1 for an overview).

7.2.3.1   Schema Level

The schema level considers mappings of general concepts of objects, like for example mapping the concept *person* in repository A to the concept *people* in repository B. The mapping type relies on the available vocabulary and may range from equality relations to complex part-of relations, depending on the language used. Besides mapping schemas onto each other, schema integration targets the creation of one general schema out of the source schemas. In any case, the result is a shared schema across repositories.

7.2.3.2   Instance Level

The instance level considers mappings of instances of concepts or objects, like for example identifying that person A is the same as person B. While instance-level integration mostly focuses on de-duplication of single instances, more complex cases, like for example determining the type of relationships between two concrete persons, may also be estimated. In general, performance decreases with increasing relationship complexity.

7.2.3.3   Declarative-Deductive Techniques

Declarative-deductive techniques provide mappings based on complex rule sets which may take use of reasoning and/or shared vocabularies. For example, the concept *person* in schema A is the same as *people* in schema B, since both are parents to the disjunctive concepts *man* and *women*. Similarly, on the instance level, the field *name* for the concepts may be the same as the joint field *forename* and *surname* for the concept *people* in schema B. Hence, declarative rules allow to map schemas and instances onto each other. However, declarative rules may not be able to include fuzziness, like different spelling of concept names or aspects like similar structures of concepts/instances.

7.2.3.4   Statistical-Inductive Techniques

Statistical-inductive techniques account the need for fuzzier matching criteria and the capability to learn from example instances. For example, de-duplication of instances – like identifying the set of unique persons from a set of persons – can be solved by clustering instances according to some similarity measures. The identified clusters constitute the unique persons. Similarly, near duplicate search can be used to match a given instance to a set of unique instances, as for example in determining identical web pages during crawling.

Along both dimensions – i.e., levels and techniques – visual analytics can support the integration process by visualizing effects of declarative rules or by visualizing the relation of certain instances to each other. Visual feedback methods allow to steer the integration process. Granitzer et al. [16] present an overview on such visually supported semantic integration processes.

7.2.4 Selection and Aggregation

Semantic enrichment and integration prepare the underlying data set for further processing. In order to reduce the number of documents subject to visual analysis, the next step includes selecting a proper subset and/or aggregating multiple documents to one single object.

7.2.4.1 Retrieval

Retrieval techniques perform the step of selecting appropriate subsets of documents. Besides the capability to search for information in text and metadata, the scalability and performance of modern retrieval techniques [34] enables feature-based filtering, query-by-example and facetted browsing, with the goal to cover all relevant documents for subsequent analysis. Aggregation and visualization methods can be applied to analyze large results sets and drill-down to task specific aspects.

7.2.4.2 Unsupervised Machine Learning

Unsupervised machine learning, in particular clustering, determines groups of similar documents based on the assumption that documents distribute over topics [55]. Groups of documents can be represented as one single data point to reduce the amount of data-points for subsequent processing steps or to improve navigation. Especially for navigation, summarizing clusters in a human readable way becomes crucial. In general, summarizations consist of extracted keywords or a brief textual description relevant to the cluster, created via summarization methods.

7.2.4.3 Summarization

Summarization methods compute a brief descriptive summary for one more documents in the form of representative text fragments or keywords. The summaries are used as labels to represent the essence of a document or a document set. Exploiting integrated metadata and structure between documents becomes essential in order to improve summarization and keyword extraction, as shown in [29]. An overview of different summarization methods can be found in [5].

7.2.5 Visualization and Interaction

The processing steps discussed above return a set of relevant objects, including features, annotations and metadata. As a next step, suitable visual layouts have to be calculated. Text data is characterized by its high-dimensional, sparse representation, which naturally leads to the application of ordination techniques.

7.2.5.1 Ordination

Ordination is a generic term describing methods for creating layouts for high-dimensional objects based on their relationships. It can be seen as a subset of dimensionality reduction techniques [11]. Dimensionality reduction methods project the high-dimensional features into a lower-dimensional visualization space, while trying to preserve the high-dimensional relationships. High-dimensional relationships can be usually expressed by similarity or distance measures. The produced layout is suitable for visualization and exploratory analysis. Other layout generation techniques, such as graph layout methods [6], are used to create a suitable visual layout for non-vector based structures, like typed graphs, temporal processes, etc.

7.2.5.2 Interactive Visualization

Interactive visualizations form the heart of any visual analytics application. Interactive components are used to visually convey information aggregated and extracted from text, and to provide means for exploratory analysis along the lines of the visual analytics mantra: "analyze first – show the important – zoom, filter and analyze further (iteratively) – details on demand" [24]. Feedback provided by users when interacting with visual representations can be fed into the previous stages of the process in order to improve its overall performance.

Visualizations usually depend on the visualized data (e.g., set, tree, graphs) and the task at hand (e.g., topical similarity, temporal development). In Sect. 7.3 we provide a detailed overview on visualizations particularly suited for text.

7.2.6 Hypothesis Formulation and Analytics Workflow

One core difference between Information Visualization and Visual Analytics lies in the support of analytical workflows and the generation and validation of hypothesis. Both, workflows and hypotheses formulation, require support from the underlying analytical process and serve as end-point towards the manipulation of all preceding steps, like acquisition, enrichment, integration, etc.

7.2.6.1 Hypothesis Generation

The visual representation of semantics in the data usually triggers new insights. New insights result in the generation of new, potentially valid hypothesis on the underlying data. For example, showing a distribution of topics in media over time may trigger the hypothesis that two events are related to each other.

Usually hypothesis generation is done in the head of the analyst, rather than making the validated hypotheses explicit. However, hypothesis generation depends on already generated and validated hypothesis. Similar to the well known "Lost in Hyperspace" effect, where users who browse the web via hyperlinks loose their initial information need, implicit hypothesis generation bears the risk to miss important, already validated facts. Hence, hypotheses and the decisions they triggered should be made explicit within an analytical process in order to guide the user. To the best of the authors knowledge this has not been done so far, but well known mathematical models for decision making processes, like the *Analytical Hierarchical Process* (AHP) [36], could be a first starting point therefore.

7.2.6.2 Hypothesis Validation

A generated hypothesis can be verified by the user. Depending on the required manipulation of the underlying analytical process, validation may range from simple interactions with the visual representation to crawling a completely new data set. For example, to see that two events are related to each other one could simply select data points related to both events and reveal their topical dependency. Hence, for efficient support of analytical tasks flexible, powerful and easy to use *task specific workflows* become important.

7.3 Visual Representations and Interactions on Text

Tight integration of visual methods with automatic techniques provides important advantages. To name a few: (i) flexibility to interchangeably apply visual and automatic techniques, as needed by the user, (ii) results of automatic text analysis, such as extracted metadata or aggregated structures, open the way for applying a wider variety of visualization techniques, which are not targeted exclusively to text, and (iii) user feedback can be used to adjust and improve the models used by automatic methods. Therefore, in this section we discuss visual representations, which primarily target textual data, and also describe how visualizations, which do not specifically target text, can be used when information is extracted from text using methods described in the previous section.

7.3.1 Topical Overview

Gaining an overview of important topics in a document set, and understanding the relationships between these topics, is crucial when users are dealing with large text repositories they are unfamiliar with. *Tag clouds* and *information landscape* are examples of visual representations which are designed to address these requirements.

7.3.1.1 Tag Clouds

Tag clouds are a popular Web 2.0 visual representation consisting of terms or short phrases which describe the content of a document or collection. Typically, keywords or named entities (such as persons or organizations) are displayed, which were extracted from document content using natural language processing methods (see Sect. 7.2.2). Size, color and layout of the words are driven by their importance, as well as by aesthetic and usability criteria [44].

In Fig. 7.3 a search result set is visualized by multiple tag clouds combined into one visualization. Each tag cloud corresponds to one of the (pre-defined) categories "sports", "politics", "europe", "society", "culture". The central tag cloud represents all documents of all categories. Each tag cloud shows the most important named entities (persons, dates, locations) for the respective category, thus giving an overview over the documents within. The polygonal boundaries for each tag cloud are generated by applying Voronoi subdivision. The initial points for generating this subdivision can either be set manually (as in the example figure) or can be the result of a similarity layout of the category content (for an example, see [46]).

7.3.1.2 Information Landscapes

Information landscapes, such as In-SPIRE [27] and InfoSky [1], employ a geographic landscape metaphor for topical analysis of large document sets. Information landscapes are primarily used for gaining an overview and for providing explorative navigation possibilities. A user who is unfamiliar with the data set is empowered to gain insight deep into the topical structure of the data, understand importance of various topics, and learn about relationships between them. As opposed to searching using queries, guided explorative navigation provides the possibility to identify interesting information even when the user's goals are vaguely defined.

In information landscapes documents are visualized as dots (icons), which are laid out in such a way that similar items are positioned close together, while dissimilar ones are placed far apart. Hills emerge where density of topically related documents is high, indicating a topical cluster. Clusters are labeled by summaries of the underlying documents, allowing users to identify areas of interest and eliminate outliers. The height of a hill is an indicator for the number of documents and the

Fig. 7.3 A visualization showing a search result set as a combination of tag clouds. Each polygonal area corresponds to a category of the documents in the search result set. Displayed named entities are enhanced with symbols indicating their type (person, location, data)

compactness of the hill is an indicator of cluster's topical cohesion. Topically similar clusters can be identified as they will appear spatially close to each other, while dissimilar clusters are separated by larger areas, visualized as sea. Aggregation of the data set and its projection into the 2D space are computed using scalable clustering and ordination algorithms, as for example described in [30, 38] (also see Sects. 7.2.4 and 7.2.5). Advanced information landscape implementations can handle data sets with far over a million documents. For such massive data sets information retrieval techniques (see Sect. 7.2.4) can be used to provide fast filtering and highlighting functionality.

Figure 7.4 shows navigation in an information landscape along a hierarchy of topical clusters, which are visualized as nested polygonal areas. Cluster labels provide a summary of the content of the underlaying documents and serve as guidance for exploration. Following the labels on each level of the hierarchy, the user can navigate the topical structure of the data and understand how clusters relate in terms of topical similarity and size. On the top-left of the figure, an overview of approximately 6,000 news articles on "computer industry" can be seen,

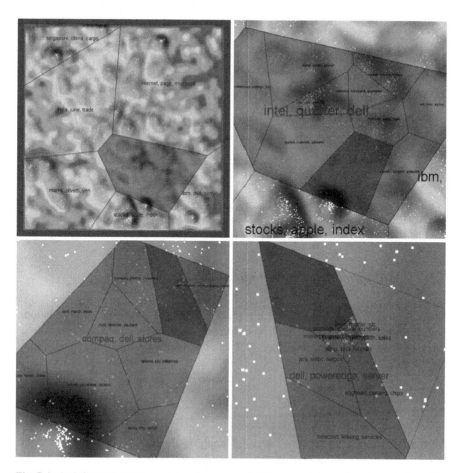

Fig. 7.4 An information landscape showing approx. 6,000 news articles on "computer industry" is used for drilling down to documents of interest: beginning with an overview (*left*) the user narrows down using topical cluster labels (*right*)

subdivided into 7 topical clusters. Clicking on the label "intel, quarter, dell", the corresponding cluster is zoomed in and the sub-areas, corresponding to its sub-clusters, are shown (top-right). Clicking on "compaq, dell, stores" (bottom-left) and then on "dell, poweredge, server" (bottom-right) narrows further down to the potential topic of interest. The cluster "poweredge, server, prices" (bottom-right) contains only five document, which can be inspected manually by the user. Free navigation by zooming (mouse-wheel) and panning (mouse-drag) is also available. Selection of documents can be performed cluster-wise, individually or on arbitrary subsets using a lasso tool.

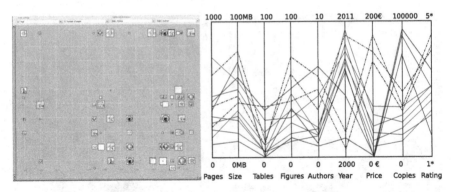

Fig. 7.5 Multidimensional visualization for books. *Left*: Scatterplot visualizing publication year (*x*-axis), page count (*y*-axis), file size (icon size), author (icon type); *right*: parallel coordinates showing nine metadata types on parallel axes

7.3.2   Multidimensional Metadata

Visualization of multidimensional metadata enables the discovery of correlations between document metadata. Such metadata may include document size and source, relevance to a search query (see Sect. 7.2.4), or extracted persons and organizations (see Sect. 7.2.2).

7.3.2.1   Scatterplot

Scatterplot is a visual representation for analysis of multidimensional metadata, mapping up to five different metadata types (dimensions) to the *x* and *y* axes, and to visual properties (color, size, shape) of displayed items [21]. The main drawback of a scatterplot is that it can correlate only a limited amount of dimensions.

7.3.2.2   Parallel Coordinates

The parallel coordinates representation [19] can handle a larger amount of dimensions, which are displayed as parallel vertical axes. For each document, the variables are displayed on their corresponding axes and connected with a polygonal line, so that patterns can be spotted easily as lines having similar shapes. In addition, a selected discrete property (e.g., class membership) can be mapped to the line color to allow identification of differentiating features for different values of the property.

Figure 7.5 (left) shows a scatterplot displaying book metadata (publication year, page count, file size, author). The scatterplot component builds upon the

Prefuse Information Visualization Toolkit[1] adding the capability to handle multiple coordinated scatterplot views (see Sect. 7.3.6 for more information on multiple coordinated views). By visualizing the same data set in two or more coordinated scatterplots at the same time, it becomes possible to increase the number of visualized dimensions above the typical five. A parallel coordinates visualization in Fig. 7.5 (right), shows nine different types of metadata for e-books, with the line style differentiating between publishers. It can be seen that the some e-books have high ratings and high prices (dashed-dotted lines), some others are cheaper and have lower ratings (continuous lines), while the remaining e-books are free and achieve highest delivery rates (dotted lines).

7.3.3 Space and Time

Visualization of geo-spatial and temporal information is very important in many applications. In what follows, we explain different approaches for producing such visualizations and discuss ThemeRiver [17], a well-known visualization conveying topical changes in large text repositories.

7.3.3.1 Visualization of Geo-Spatial Information

The visualization of geo-spatial information, as for example extracted locations, is a natural fit for the application of various geo-visualization approaches [7]. A popular application of geo-visualization is to show automatically extracted spatial information (see Sect. 7.2.2) on geographical maps [39] in order to reveal where something is happening. Figure 7.6 shows a geo-spatial visualization of locations extracted from German news articles [28]. The extracted locations are depicted on a map of Austria as cones, where the size of a cone corresponds to the number of news articles the location occurred in. Clicking on a cone triggers a filtering of the news article set by the selected location, and thus this visualization can be used as a faceted search tool.

Geo-spatial visualizations are not restricted to geographic maps; they can also be applied in e.g., virtual 3D environments. An example is the planetarium that has been integrated into an encyclopedia application [26], providing coordination between browsing spatial (astronomic) references in text and navigation in the virtual environment.

---

[1]http://prefuse.org/.

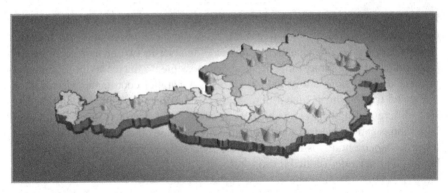

Fig. 7.6 Geo-visualization of Austria showing geo-references in news articles (*cones*). The size of the cone corresponds to the number of news articles for the particular geo-reference

7.3.3.2 Visualization of Temporal Information

The visualization of temporal information, such as document creation date or automatically extracted time references (see Sect. 7.2.2), can be realized by a variety of visual components. Although different in many aspects, visual representations for temporal data usually share a common feature: they include a visual element which symbolizes the flow of time. For example, temporal data can be visualized along a straight line or along a spiral [54], both representing the flow of time. Although a straight time axis is more common, a spiral time axis has the advantage of being suitable for detecting cycles and recurring events, and it allows for displaying long time intervals with high temporal resolution even on small screens.

7.3.3.3 ThemeRiver

ThemeRiver [17] is a well-known visualization conveying topical changes in large text repositories. It uses a metaphor of flowing river streams to visualize trends and changes in topical clusters, in the context of external events (see clustering in Sect. 7.2.4). In addition to topical clusters, metadata clusters, for example documents mentioning a specific location, can also be visualized. ThemeRiver empowers users not only to understand trends but also to discover correlations and causal relationships between clusters.

In Fig. 7.7 a stream visualization, which closely resembles the ThemeRiver, shows temporal development of topical clusters for approximately 750 news documents on "oil spill". The x-axis symbolizes the flow of time, while the y-axis conveys the amount of documents at a given moment in time. Each topical cluster is represented by a stream of particular color, where the width of the stream along the time axis correlates with the number of documents. By observing the development of the "japan, tokyo, bay" topical cluster (second from bottom), which has two distinctive peaks, it is obvious that temporal development of the "russia" metadata cluster (bottom-most) correlates with the first peak, but not with the second.

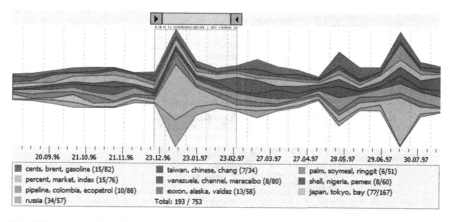

Fig. 7.7 A stream visualization of approx. 750 news documents on "oil spill", showing temporal development. Different gray values correspond to different topics

Naturally, a fusion of both spatial and temporal information in one visualization also leads to interesting results. For example, the three-dimensional GeoTime [22] visualization depicts a geographic map where the flow of time is orthogonal to the map (i.e. on the z-axis). In this way GeoTime facilitates tracking of ground movements over time and identification of activity hot-spots in both space and time.

7.3.4 Relationships

Relationships between concepts (e.g., keywords or named entities), identified by methods such as co-occurrence analysis and disambiguation techniques (see Sect. 7.2.3), are typically presented using graph visualizations [18]. For example, PhraseNet [53] displays relationships between terms within a document, while FacetAtlas [4] relies on faceted retrieval to visualize relationships between faceted metadata. Relationships between aggregated structures (see Sect. 7.2.4), such as document clusters, can be visualized by Cluster Maps [10]. It is a representation similar to Venn and Euler diagrams, showing whether (and through which features) different clusters overlap topically.

7.3.4.1 Graph Visualization

A graph visualization that is used to present relationships extracted from approximately 25,000 documents can be seen in Fig. 7.8. Concepts (keywords) are placed in the 2D plane, depending on their interconnectedness, using a force-directed placement method (see Sect. 7.2.5). An edge bundling technique [25] is applied to reduce clutter, which would otherwise occur due to the high number of relationships.

Fig. 7.8 A graph visualization of relationships between concepts extracted from a text data set (data courtesy of German National Library of Economics, 2011). Note that edge bundling is used to improve clarity and reduce clutter in the edge layout

To preserve clarity even when visualizing larger graphs, a level-of-detail sensitive algorithm decides which informations is displayed and which is hidden depending on user focus and the current zoom level. To navigate, the user clicks on a concept which triggers a zoom-in operation focusing that concept. Concepts close to the chosen one are displayed in more details, revealing finer structures in the graph.

7.3.5 Visually Enhanced User Feedback

Analytical tasks require well-designed interaction mechanisms along with different kinds of visualizations. Interactions can be grouped along three orthogonal dimensions, namely (i) the kind of *operation* they perform, i.e., navigation, selection and manipulation, (ii) the *modality of the interaction*, i.e., query, point-and-click, language input, multi-touch, and (iii) its *influence on the underlying analytical process*, i.e., the adaptation of data, parameters or the mining models themselves. This subsection will briefly discuss (i) and (ii), and then focus on how interactions can be used to steer the underlying analytical process.

7.3.5.1 Modalities of Interaction

Modalities of interaction depend mostly on the input devices. A search box can be seen as "textual modality" which allows to filter relevant documents based on keywords, provided either via keyboard or speech-to-text. Clearly, with the advance of multi-touch devices new capabilities in expressing user needs become available. While modalities of interaction influence the design of visualization and determine how interactions take place, they do not influence the possible operations on the data and the steering capabilities on the analytics process. For a detailed overview of different interaction modalities the reader is referred to [49].

7.3.5.2 Operations of Interactions

Operations of interactions describe the purpose of an interaction. Interactions to navigate complex information spaces, to drill down on particular interesting patterns and to switch between different perspectives, are the most common navigational operations. Examples are browsing hyper-links or navigating a hierarchical structure (see Fig. 7.4, left). Selection comprises operations that allow users to select data points of interest and their properties. Examples include multi-selection in a list of documents or lasso selection of data points in a similarity layout of documents (see Fig. 7.4, top-right). Finally, manipulations form the essence of any visual analytics application. Manipulative interactions, like removing certain data points from the analysis or assigning a group of documents to a particular class, allow to steer the underlying classification, clustering and retrieval processes.

7.3.5.3 Steering the Visual Analytics Process

Given interactions of different modalities and operations, the question remains how the underlying process could be steered. In the following, we will discuss steering on the *data-point level* and the *model level*.

Data-point level: On the data-point level, selecting a particular subset of data points or a subset of data sources for the detailed analysis becomes the most common form of influence. For example, Fig. 7.9 (left) shows a similarity layout comparing search results from different search engines [51]. Sources could be interactively added or removed in order to change the topical layout. Similarly, in the landscape visualization shown in Fig. 7.4, a lasso selection can be used to select a set of similar documents. This set could then be used as a positive set of examples for training supervised classification algorithm.

Model level: On the model level, the goal is to control the underlying mining model. The most direct form of controlling mining models is by setting parameters directly, for example, the number of clusters or cost functions for positive or negative classification errors. However, comprehension of resulting effects of direct parameter manipulation becomes non-trivial especially for data-mining laymen. Hence, we propose "direct manipulations" of mining models using visualizations.

The concept of direct manipulation greatly improved user interfaces of computers by allowing users to directly manipulate information objects, like files and folders.

Users have been empowered to drag and drop object instead of manipulating them via a command line. In analogy, we give two examples on direct manipulation in visual analytics.

Example 1. As a first example, consider again the landscape visualization shown in Fig. 7.4. High-dimensional data points have been projected onto the 2D-plane using clustering and ordination techniques. However, the high-dimensional distance measure, and therefore the result of the projection, may not fit to the users expectation of "distance" between documents. Some topics may be too close and some too far away from each other. Instead of directly changing the distance function or parameters of the ordination technique, the user could directly drag topically similar data points closer to each other and dissimilar data-points further apart, yielding user-determined distances between data points. By applying metric learning techniques [14, 47] the user-determined similarity could be transformed into a high-dimensional distance function in some kind of inverse projection [15]. The resulting high-dimensional distance function can be used in different mining algorithms to reflect the user's notion of "similarity" between documents.

Example 2. A second example concerns supervised machine learning models. Interactions on a visual layout may be used two-fold: (i) to correct classification errors or re-force correct classifications, and (ii) to generate new training data. A visualization supporting these tasks requires the following properties: First, the visualization should allow to judge problematic behavior of classification models, like biases towards particular classes. Second, fast and easy identification of false and/or problematic examples, e.g., outliers should be supported. Third, users should be able to rapidly select and (re-)label examples. Further, if is preferable to have the same kind of visualization, independently of the classification task and the employed data classification algorithm.

The visualization proposed in [42, 43] satisfies these properties and has been shown to support users in improving classification models [41]. Here, classes are arranged around a circle. Data-points are placed in the interior of the circle with their distances to every class being proportional to the a-posterior probability that a data point belongs to that class (see Fig. 7.9, right). Data points can be inspected, selected and dragged to the correct class resulting in re-training and improving the underlying text classifier. A combined user interface employing this visualization and an information landscape has further been shown applicable to generate classifier hypothesis from scratch [45].

7.3.6 Multiple Visualization Interfaces

Complex analytic scenarios involve heterogeneous data repositories consisting of different types of information. Visual representations are designed to target specific aspects of the data, such as metadata correlations, topical similarity, temporal

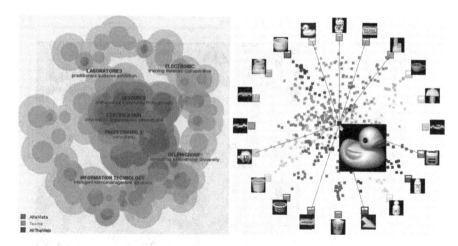

Fig. 7.9 Examples for visually enhanced feedback. *Left*: Search results (*circles*) for comparing the topic overlap of different search engines (colors). Results with similar content are close. Sources can be interactively added or removed. *Right*: Visualizing classification decisions. Classes are arranged in a circle, data points are placed inside the circle according to their a-posteriori probabilities. Decisions can be corrected by drag and drop, classifier is retrained

developments, geo-locations, etc. When simultaneous analysis of different information types is required, user interfaces consisting of multiple visual components are necessary. One way to address visual analysis of heterogeneous data is to integrate various visualizations within a single immersive 3D virtual environment, such as the Starlight System [35]. A more widely used approach is *Coordinated Multiple Views* (CMV) [31]. Multiple view coordination is a technique for tightly coupling multiple visualization components into a single coherent user interface, so that changes triggered by interactions in one component are immediately reflected in all others components.

Figure 7.10 shows a coordinated user interface, consisting of an information landscape, a stream visualization, as well as of several other widgets, such as trees and tables. The interface is used for "fused" analysis of topical, temporal and metadata aspects of large text repositories [37]. The tree component, on the left, shows the hierarchy of topical clusters providing a virtual table of contents. An information landscape (see Sect. 7.3.1), on the right, visualizes document frequency and topical similarity of clusters and documents. A stream view (see Sect. 7.3.3), on the bottom, conveys temporal development of topical (and metadata) clusters. Two additional components are available but are hidden in the screenshot: a faceted metadata tree showing extracted persons, organizations and locations, and a table providing detailed information on clusters and documents.

The coordination of components includes the following: (i) *navigation in the cluster hierarchy* (triggered in any of the components), (ii) *document selection*

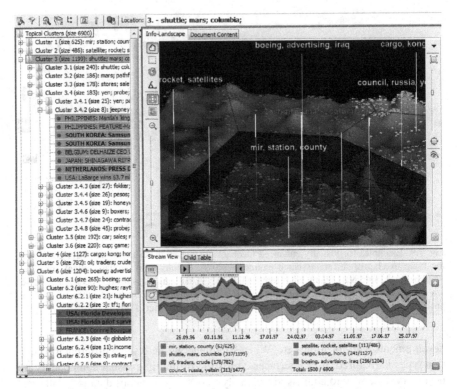

Fig. 7.10 A coordinated multiple views GUI showing 6,900 news documents on "space". Document selection (by time: from June to August), document coloring (each topical cluster in different color) and navigation in the hierarchy (location: Cluster 3 "shuttle, mars, columbia") are coordinated

(lasso-selection in the landscape, temporal selection in stream view, or cluster-wise selection in the trees), (iii) *document coloring* (driven by the stream view color assignments), and (iv) *document icons* (user-assignable from any component).

Coordination ensures that all views will focus on the same cluster, and that document selection, colors and icons are consistent in all views. In this way, discovery of patterns over the boundaries of individual visualizations becomes possible. For example, topical-temporal analysis can be performed by selecting documents belonging to two temporally separate events in the stream view, and then inspecting in the landscape whether those documents are topically related or not. Moreover, correlations between topical clusters and occurrences of a metadatum (e.g., persons) can be identified by assigning different icons to documents mentioning different persons, and then observing the distribution of these persons over topical clusters in the landscape.

7.4   Application Scenarios and Domains

Application scenarios for visual analysis and discovery in text repositories can be identified in a wide range of domains. News media, encyclopedia volumes, scientific paper repositories, patent databases or intelligence information systems, represent an exemplary selection of domains to which the methods discussed in this chapter have been beneficially applied. Given the diversity of application scenarios, we will try to impose some structuring along relevant dimensions.

A first important dimension involves the target user group of a given application. Clearly, the skill and experience level of the expected target user group should influence the choice of visual means. *Information Visualization* and *Visual Analytics* approaches usually focus on efficiency. This results in visual means which are perfectly suitable for expert analysts, who have a high level of visual literacy and domain knowledge. In contrast, *Knowledge Visualization* approaches [8] focus on comprehensibility. The resulting visual means are often less efficient and flexible, but can be utilized by a general audience.

A second dimension considers the amount of a priori information and context available in a given application scenario. If information or context is available, for example in the form of a formulated query or user profile information, an initial search can limit the number of information items which have to be considered. In this case, the visual analysis and manipulation of search results becomes the prevalent task. In the absence of explicit information or context, explorative visualizations can enable the discovery of facts without having to explicate an information need in advance. The following application scenarios provide a representative cross section along this dimensions.

7.4.1   Media Analysis for the General Public

Media Analysis providers have traditionally shaped their services towards the requirements of decision makers in enterprises and organizations. The advent of the World Wide Web and the introduction of consumer-generated media has greatly increased the amount of news sources available to a general audience. Media consumers today find themselves assuming the role of media analysts in order to satisfy personal information needs. News visualization has been a favored use case for Information Visualization almost from the beginnings of this discipline [33]. However, in the spirit of the structure established above, visual support for this application scenario should employ simple visual means and assume limited visual literacy.

The *Austrian Press Agency* (APA) has provided a general audience with a number of experimental news visualizations through its labs platform since 2008 [28]. From a technical point of view, the platform implements the pipeline architecture outlined in this chapter. The acquisition stage relies on the PowerSearch media

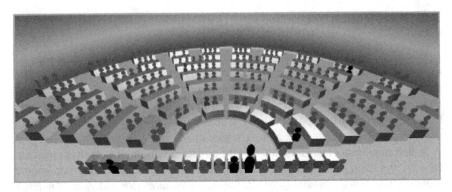

Fig. 7.11 A visualization of occurrences of Austrian politicians in search results. A rendered model of the parliament is used as visual metaphor. The figures of politicians are colored in their party color and scaled relative to the occurrence count. Clicking on a figure narrows the search result to articles containing the selected politician

database run by the company, which provides 180 Million news articles from 250 sources in a normalized manner. Semantic enrichment is facilitated through a combination of rule-based and dictionary-based methods, which annotate persons, locations and companies. Machine learning techniques are used to classify articles into topical areas. Semantic integration is currently being addressed, for instance by harmonizing identified persons with appropriate data sources from encyclopedias. Retrieval is performed through a classical query-based interface, which provides relevance-ranked search result lists.

The initial architecture has been tailored towards faceted filtering of large search result sets. Given a query entered by a user, the system generates the result set and displays a variety of visualizations, each of which represents a certain facet. For instance, the occurrence of members of parliament and members of government in a set of search results is visualized in a model of the Austrian parliament, as shown in Fig. 7.11. Other visualizations include a geo-spatial view, a round table view of prominent politicians and a tag cloud. All visualizations are very simple in design, rely on metaphors to ease understanding and support a very simple interaction scheme: Selecting a visual entity filters the result set to results containing the entity. Experiments have shown that this kind of system is accessible to a general audience without training.

An example visualization of more complex media analysis results is shown in Fig. 7.12. Co-occurrence of key political figures extracted from a text corpus is represented using a node-link-diagram in which links have been bundled to reveal high-level patterns [25]. This kind of visualization favors an exploratory approach which reveals general trends of the whole article set in the absence of a concrete search query.

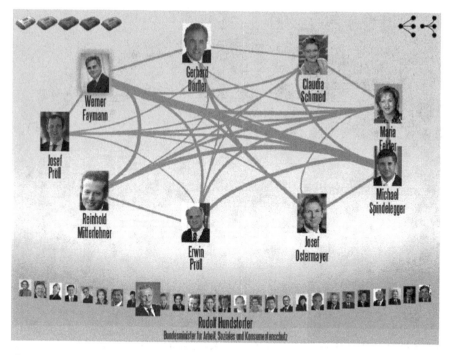

Fig. 7.12 A visualization of co-occurrences of Austrian politicians in recent news media. Politicians are displayed as nodes connected by links representing co-occurrence strength by line width. Links are bundled to reveal high-level edge patterns. The strongest link visible is between the chancellor and the vice-chancellor

7.4.2 Navigation and Exploration of Encyclopedias

Modern digital encyclopedias contain hundreds of thousands of textual articles and multimedia elements, which constitute a knowledge space encompassing virtually all areas of general interest. Traditional retrieval and discovery techniques in this domain have included keyword search for articles and cross-reference based navigation between articles. The German-language Brockhaus encyclopedia provides a visualization system which enables the visual navigation of article context. This three-dimensional Knowledge Space visualization presents topically related articles, using figurative graphical elements as visual metaphors. The idea behind the visualization is to support navigation between articles and to encourage exploration of the encyclopedia in the spirit of edutainment.

The visualization shown in Fig. 7.13, displays the currently selected article at the center of a disc divided into topical segments and arranges similar articles around it. Relevant articles are placed close to the center and each article is placed within the segment corresponding to its topic (chosen from a ten-item topic scheme). Articles are represented by shapes according to type: *cylinders* represent

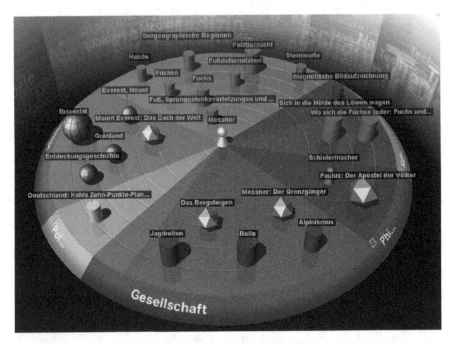

Fig. 7.13 The "Knowledge Space" visualization displaying the context of the encyclopedia entry for the mountaineer Reinhold Messner (*center*). The disc is divided into segments representing topics (e.g., "society" in the front). Related articles are represented by objects placed on the disc; shape, size and color encode additional metadata. For example, in the leftmost segment a geographic article (*circle*) and a premium content article (*diamond*) about the Mountain Everest is shown

factual articles, *spheres* represent geographic articles, *cones* represent biographic articles and *diamonds* represent articles featuring premium content. Article labels are displayed above the shapes. Dragging the mouse horizontally spins the disc around its central axis. Dragging the mouse vertically adjusts zoom factor and vertical view angle. Clicking on an object navigates to the corresponding article.

7.4.3 Patent Analysis and Comparison

The identification of prior art, and the discovery of patterns and trends in patents constitutes a crucial aspect of business intelligence for innovative enterprises. The raw data for this kind of analysis is readily available in the form of various commercial and open patent databases. However, the actual information contained in patents is very hard to analyze and understand. This phenomenon stems, in part, from deliberate attempts to paraphrase key issues in order to maintain a competitive advantage. Another reason for the complexity of patent information is the huge amount of domain knowledge required to make sense of an actual patent, covering a narrow technical aspect.

The Austrian company *m2n* has created a patent analysis system which has been used by various large enterprises, for instance by one of the largest global steel manufacturers. This system displays patent data sets, acquired from a number of configurable sources, in a multiple coordinated view environment, which integrates textual and visual representations [37]. The visualization application includes an information landscape, a temporal visualization and a number of other coordinated views, similar to the user interface shown in Fig. 7.10. Referring to the structure established above, this system clearly targets expert users which accept a large amount of training in order to harvest all the benefits.

7.5 Conclusion and Outlook

Through combining visually supported reasoning with large scale automatic processing, visual analytics opens new possibilities for exploration and discovery of knowledge in text repositories. Aggregation and summarization are central to scaling visualizations to very large data sets. Retrieval techniques enable filtering, highlighting and selection on repositories of virtually unlimited size. Information extraction opens the way for using visual representations which are not directly related to text, such as geo-visualization or graph visualization. Finally, visualization not only introduces human knowledge and visual pattern recognition into the analytical process, but also provides the possibility to improve the performance of automatic methods through consideration of user feedback.

While it is hard to deliver predictions on future development of the field, the following directions appear promising: Triggered by the surge in use of smart mobile devices and multi-touch interfaces, support for collaborative scenarios using new input devices, such as tablets and multi-touch tables, is likely to gain traction. On the algorithm side, the peculiarities of the emerging phenomenon of social networks and social media, such as quality and trustworthiness of information, pose new challenges. In the quest to handle ever larger data sets the efficient exploitation of the cloud for computation and storage holds the promise of ultimate scalability.

References

1. Andrews, K., Kienreich, W., Sabol, V., Becker, J., Droschl, G., Kappe, F., Granitzer, M., Auer, P., Tochtermann, K.: The infoSky visual explorer: exploiting hierarchical structure and document similarities. Inf. Vis. **1**(3–4), 166–181 (2002)
2. Bleiholder, J., Naumann, F.: Data fusion. ACM Comput. Surv. **41**, 1:1–1:41 (2009)
3. Bruijn, J.d., Ehrig, M., Feier, C., Martìns-Recuerda, F., Scharffe, F., Weiten, M.: Ontology Mediation, Merging, and Aligning, in Semantic Web Technologies: Trends and Research in Ontology-based Systems (eds J. Davies, R. Studer and P. Warren), John Wiley & Sons, Ltd, Chichester, UK. pp. 95–113. (2006). doi:10.1002/047003033X.ch6

4. Cao, N., Sun, J., Lin, Y.R., Gotz, D., Liu, S., Qu, H.: Facetatlas: multifaceted visualization for rich text corpora. IEEE Trans. Vis. Comput. Graph. **16**(6), 1172–1181 (2010)
5. Das, D., Martins, A.F.: A survey on automatic text summarization. Technical report, Carnegie Mellon University (2007). Literature Survey for the Language and Statistics II course at CMU
6. Díaz, J., Petit, J., Serna, M.: A survey of graph layout problems. ACM Comput. Surv. **34**, 313–356 (2002)
7. Dykes, J., MacEachren, A.M., Kraak, M.J. (eds.): Exploring Geovisualization. Elsevier, Amsterdam (2005)
8. Eppler, M.J., Burkhard, R.A.: Knowledge visualization. In: Schwartz, D. & D. Te'eni (eds.) Encyclopedia of Knowledge Management, Second Edition, PA: Information Science Reference. pp. 987–999. Hershey. doi:10.4018/978-1-59904-931-1.ch094
9. Fayyad, U.M., Piatetsky-Shapiro, G., Smyth, P.: From data mining to knowledge discovery in databases. AI Mag. **17**, 37–54 (1996)
10. Fluit, C.: Autofocus: semantic search for the desktop. Inf. Vis. Int. Conf. **0**, 480–487 (2005)
11. Fodor, I.: A survey of dimension reduction techniques. Technical report UCRL-ID-148494, US DOE Office of Scientific and Technical Information (2002)
12. Gantz, J.F., Reinsel, D., Chute, C., Schlichting, W., McArthur, J., Minton, S., Xheneti, I., Toncheva, A., Manfrediz, A.: The expanding digital universe, a forecast of worldwide information growth through 2010. IDC White Paper – sponsored by EMC (2007)
13. Gantz, J.F., Chute, C., Manfrediz, A., Minton, S., Reinsel, D., Schlichting, W., Toncheva, A.: The diverse and exploding digital universe, an updated forecast of worldwide information growth through 2011. IDC White Paper – sponsored by EMC (2008)
14. Granitzer, M.: Adaptive term weighting through stochastic optimization. In: Gelbukh, A. (ed.) Computational Linguistics and Intelligent Text Processing. Lecture Notes in Computer Science, vol. 6008, pp. 614–626. Springer, Berlin/Heidelberg (2010)
15. Granitzer, M., Neidhart, T., Lux, M.: Learning term spaces based on visual feedback. In: International Workshop on Database and Expert Systems Applications (DEXA), Krakow, pp. 176–180. IEEE Computer Society (2006)
16. Granitzer, M., Sabol, V., Onn, K.W., Lukose, D., Tochtermann, K.: Ontology alignment – a survey with focus on visually supported semi-automatic techniques. Future Internet **2**(3), 238–258 (2010)
17. Havre, S., Hetzler, E., Whitney, P., Nowell, L.: ThemeRiver: visualizing thematic changes in large document collections. IEEE Trans. Vis. Comput. Graph. **8**(1), 9–20 (2002)
18. Herman, I., Melançon, G., Marshall, M.S.: Graph visualization and navigation in information visualization: A survey. IEEE Trans. Vis. Comput. Graph. **6**, 24–43 (2000)
19. Inselberg, A., Dimsdale, B.: Parallel coordinates for visualizing multi-dimensional geometry. In: CG International '87 on Computer Graphics 1987. Springer-Verlag New York, Inc., Karuizawa, Japan, New York, NY, USA, pp. 25–44 (1987). http://dl.acm.org/citation.cfm?id=30300.30303
20. Kaiser, K., Miksch, S.: Information extraction – a survey. Technical report Asgaard-TR-2005-6, Vienna University of Technology (2005)
21. Kandlhofer, M.: Einbindung neuer Visualisierungskomponenten in ein Multiple Coordinated Views Framework, Endbericht Master-Praktikum (2008)
22. Kapler, T., Wright, W.: Geo time information visualization. Inf. Vis. **4**, 136–146 (2005)
23. Keim, D.A., Mansmann, F., Oelke, D., Ziegler, H.: Visual analytics: combining automated discovery with interactive visualizations. In: Discovery Science, LNAI, Springer Berlin/Heidelberg, Budapest, Hungary, pp. 2–14 (2008)
24. Keim, D.A., Mansmann, F., Schneidewind, J., Thomas, J., Ziegler, H.: Visual analytics: scope and challenges. In: Simoff, S.J., Böhlen, M.H., Mazeika, A. (eds.) Visual Data Mining, pp. 76–90. Springer, Berlin/Heidelberg (2008)
25. Kienreich, W., Seifert, C.: An application of edge bundling techniques to the visualization of media analysis results. In: Proceedings of the International Conference on Information Visualization, London. IEEE Computer Society Press (2010)

26. Kienreich, W., Zechner, M., Sabol, V.: Comprehensive astronomical visualization for a multi-media encyclopedia. In: International Symposium of Knowledge and Argument Visualization; Proceedings of the International Conference Information Visualisation, Zurich, pp. 363–368. IEEE Computer Society (2007)

27. Krishnan, M., Bohn, S., Cowley, W., Crow, V., Nieplocha, J.: Scalable visual analytics of massive textual datasets. In: IEEE International Parallel and Distributed Processing Symposium, 2007. IPDPS 2007, Long Beach, pp. 1–10 (2007)

28. Lex, E., Seifert, C., Kienreich, W., Granitzer, M.: A generic framework for visualizing the news article domain and its application to real-world data. J. Digit. Inf. Manag. **6**, 434–441 (2008)

29. Muhr, M., Kern, R., Granitzer, M.: Analysis of structural relationships for hierarchical cluster labeling. In: Proceedings of the International ACM Conference on Research and Development in Information Retrieval (SIGIR), SIGIR '10, Geneva, pp. 178–185. ACM, New York (2010)

30. Muhr, M., Sabol, V., Granitzer, M.: Scalable recursive top-down hierarchical clustering approach with implicit model selection for textual data sets. In: IEEE International Workshop on Text-Based Information Retrieval; Proceedings of the International Conference on Database and Expert Systems Applications, Bilbao (2010)

31. Müller, F.: Granularity based multiple coordinated views to improve the information seeking process. Ph.D. thesis, University of Konstanz, Germany (2005)

32. Muthukrishnan, P., Radev, D., Mei, Q.: Edge weight regularization over multiple graphs for similarity learning. In: IEEE 10th International Conference on Data Mining (ICDM), 2010, Sydney, pp. 374–383 (2010). doi:10.1109/ICDM.2010.156

33. Rennison, E.: Galaxy of news: an approach to visualizing and understanding expansive news landscapes. In: Proceedings of the ACM Symposium on User Interface Software and Technology, UIST '94, Marina del Rey, pp. 3–12. ACM, New York (1994)

34. Ribeiro-Neto, B., Baeza-Yates, R.: Modern Information Retrieval: The Concepts and Technology Behind Search, 2nd edn. Pearson Education, Ltd., Harlow, England, Addison-Wesley (2011). http://dblp.uni-trier.de

35. Risch, J.S., Rex, D.B., Dowson, S.T., Walters, T.B., May, R.A., Moon, B.D.: The STARLIGHT information visualization system. Readings in Information Visualization, pp. 551–560. Morgan Kaufmann, San Francisco (1999)

36. Saaty, T.L.: Principia Mathematica Decernendi: Mathematical Principles of Decision Making, 1st edn. RWS Publications, Pittsburgh, PA, USA (2010)

37. Sabol, V., Kienreich, W., Muhr, M., Klieber, W., Granitzer, M.: Visual knowledge discovery in dynamic enterprise text repositories. In: Proceedings of the International Conference Information Visualisation (IV), pp. 361–368. IEEE Computer Society, Washington, DC (2009)

38. Sabol, V., Syed, K., Scharl, A., Muhr, M., Hubmann-Haidvogel, A.: Incremental computation of information landscapes for dynamic web interfaces. In: Proceedings of the Brazilian Symposium on Human Factors in Computer Systems, Barcelona, Belo Horizonte, Brazil pp. 205–208 (2010). http://dblp.uni-trier.de/db/conf/ihc/ihc2010.html#SabolSSMH10

39. Scharl, A., Tochtermann, K.: The Geospatial Web: How Geobrowsers, Social Software and the Web 2.0 are Shaping the Network Society (Advanced Information and Knowledge Processing). Springer, New York/Secaucus (2007)

40. Sebastiani, F.: Machine learning in automated text categorization. ACM Comput. Surv. **34**(1), 1–47 (2002)

41. Seifert, C., Granitzer, M.: User-based active learning. In: Fan, W., Hsu, W., Webb, G.I., Liu, B., Zhang, C., Gunopulos, D., Wu, X. (eds.) Proceedings of the International Conference on Data Mining Workshops (ICDM), Sydney, pp. 418–425 (2010)

42. Seifert, C., Lex, E.: A novel visualization approach for data-mining-related classification. In: Proceedings if the International Conference on Information Visualisation (IV), Barcelona, pp. 490–495. Wiley (2009)

43. Seifert, C., Lex, E.: A visualization to investigate and give feedback to classifiers. In: Proceedings of the European Conference on Visualization (EuroVis), Berlin (2009). Poster

44. Seifert, C., Kump, B., Kienreich, W., Granitzer, G., Granitzer, M.: On the beauty and usability of tag clouds. In: Proceedings of the International Conference on Information Visualisation (IV), London, pp. 17–25. IEEE Computer Society, Los Alamitos (2008)
45. Seifert, C., Sabol, V., Granitzer, M.: Classifier hypothesis generation using visual analysis methods. In: Zavoral, F., Yaghob, J., Pichappan, P., El-Qawasmeh, E. (eds.) Networked Digital Technologies. Communications in Computer and Information Science, vol. 87, pp. 98–111. Springer, Berlin/Heidelberg (2010)
46. Seifert, C., Kienreich, W., Granitzer, M.: Visualizing text classification models with Voronoi word clouds. In: Proceedings of the International Conference Information Visualisation (IV), London (2011). Poster
47. Shalev-Shwartz, S., Singer, Y., Ng, A.Y.: Online and batch learning of pseudo-metrics. In: International Conference on Machine learning (ICML), Banff, p. 94 (2004)
48. Shneiderman, B.: Inventing discovery tools: combining information visualization with data mining. Inf. Vis. 1(1), 5–12 (2002)
49. Shneiderman, B., Plaisant, C.: Designing the User Interface: Strategies for Effective Human-Computer Interaction, 5th edn. Addison-Wesley Publ. Co., Reading, MA, p. 606 (2010)
50. Thomas, J.J., Cook, K.A. (eds.): Illuminating the Path: The Research and Development Agenda for Visual Analytics. IEEE Computer Society, Los Alamitos (2005)
51. Tochtermann, K., Sabol, V., Kienreich, W., Granitzer, M., Becker, J.: Enhancing environmental search engines with information landscapes. In: International Symposium on Environmental Software Systems, Semmering. http://www.isess.org/ (2003)
52. Tukey, J.W.: Exploratory Data Analysis, 1st edn. Addison Wesley, Massachusetts (1977)
53. van Ham, F., Wattenberg, M., Viegas, F.B.: Mapping text with phrase nets. IEEE Trans. Vis. Comput. Graph. 15, 1169–1176 (2009)
54. Weber, M., Alexa, M., Muller, W.: Visualizing time-series on spirals. In: IEEE Symposium on Information Visualization, 2001. INFOVIS 2001, San Diego, pp. 7–13 (2001)
55. Xu, R., Wunsch, D.: Survey of clustering algorithms. IEEE Trans. Neural Netw. 16(3), 645–678 (2005)

Chapter 8
Practical Distributed Privacy-Preserving Data Analysis at Large Scale

Yitao Duan and John Canny

Abstract In this chapter we investigate practical technologies for security and privacy in data analysis at large scale. We motivate our approach by discussing the challenges and opportunities in light of current and emerging analysis paradigms on large data sets. In particular, we present a framework for privacy-preserving distributed data analysis that is practical for many real-world applications. The framework is called *Peers for Privacy* (P4P) and features a novel heterogeneous architecture and a number of efficient tools for performing private computation and offering security at large scale. It maintains three key properties, which are essential for real-world applications: (i) provably strong privacy; (ii) adequate efficiency at reasonably large scale; and (iii) robustness against realistic adversaries. The framework gains its practicality by decomposing data mining algorithms into a sequence of vector addition steps, which can be privately evaluated using efficient cryptographic tools, namely verifiable secret sharing over *small* field (e.g., 32 or 64 bits), which have the same cost as regular, non-private arithmetic. This paradigm supports a large number of statistical learning algorithms, including SVD, PCA, k-means, ID3 and machine learning algorithms based on Expectation-Maximization, as well as all algorithms in the statistical query model (Kearns, Efficient noise-tolerant learning from statistical queries. In: STOC'93, San Diego, pp. 392–401, 1993). As a concrete example, we show how singular value decomposition, which is an extremely useful algorithm and the core of many data mining tasks, can be performed efficiently with privacy in P4P. Using real data, we demonstrate that P4P is orders of magnitude faster than other solutions.

Y. Duan (✉)
NetEase Youdao, Beijing, China
e-mail: duan@rd.netease.com

J. Canny
Computer Science Division, University of California, Berkeley, CA 94720, USA
e-mail: jfc@cs.berkeley.edu

A. Gkoulalas-Divanis and A. Labbi (eds.), *Large-Scale Data Analytics*,
DOI 10.1007/978-1-4614-9242-9_8, © Springer Science+Business Media New York 2014

8.1 Introduction

Imagine the scenario where a large group of users want to mine their collective data. This could be a community of movie fans extracting recommendations from their ratings, or a social network voting for their favorite members. In all these cases, the users may wish not to reveal their private data, not even to a "trusted" service provider, but still obtain verifiably accurate results. The major issues that make this kind of tasks challenging are the scale of the problem and the need to deal with malicious users. Typically the quality of the result increases with the size of the data (both the size of the user group and the dimensionality of per user data). Nowadays, it is common for commercial service providers to run algorithms on data sets collected from thousands or even millions of users. For example, the well-publicized Netflix Prize data set[1] consists of roughly 100 M ratings of 17,770 movies contributed by 480 K users. At such a scale, both private computation and verifying proper user behavior become very difficult (more on this later).

In this chapter we investigate practical technologies for security and privacy in data analysis at large scale. Our focus will be on a distributed setting, where private data resides on each client. Compared to a centralized setting, this is a more challenging model: due to the lack of a trusted entity, the privacy objective mandates that data must be obscured in some way (e.g., through encryption), which not only adds much cost but also complicates many issues, such as verification of various security properties. However, we believe that the principles and techniques we develop are general and can be applied to other settings as well.

Our goal is to provide a privacy solution that is practical for several real-world applications at reasonably large scale. In particular, we present a framework for such tasks. The framework, first introduced in [32], is called *Peers for Privacy* (P4P) and features a novel heterogeneous architecture and a number of efficient tools for performing private computation and ensuring security at large scale. It maintains the following properties which we believe are essential for real-world applications: (1) provably strong privacy; (2) adequate efficiency at reasonably large scale; and (3) robustness against realistic adversaries.

The proposed framework gains its practicality by decomposing data mining algorithms into a sequence of vector addition steps that can be privately evaluated using efficient cryptographic tools, namely *verifiable secret sharing* (VSS) over *small* field (e.g., 32 or 64 bits), which have the same cost as regular, non-private arithmetic. This paradigm supports a large number of statistical learning algorithms, including SVD, PCA, k-means, ID3, and machine learning algorithms based on *Expectation-Maximization* (EM), as well as all algorithms in the *statistical query model* [50]. We believe it is a promising step towards bringing practical privacy and security to large-scale data analysis applications.

---

[1]http://www.netflixprize.com/.

8.2   Background and Related Work

Privacy issues arise in all aspects of information utilization, from data collection and storage, to analysis and release. To provide a holistic solution, we draw on a large number of related works from a few areas. We survey their results in this section.

8.2.1   Cryptography

Cryptography provides powerful primitives for protecting communication and computation against all kinds of adversarial impacts. These primitives can be used to build systems with properties that are rigorously provable, either unconditionally or under some reasonable assumptions. Besides encryption, modern cryptography also provides tools, such as *secure multiparty computation* (MPC) and *zero-knowledge proof* (ZKP), which offer mechanisms for verifying security properties without violating data privacy. MPC allows n players to compute a function over their collective data. It guarantees the privacy of their inputs and the correctness of the outputs, even when some players are corrupted by the same adversary. The problem dates back to Yao [77] and Goldreich et al. [43], and has been extensively studied in cryptography (see e.g., [3, 7, 46]). In recent years we observe some significant improvement in efficiency. Some protocols achieve nearly optimal asymptotic complexity [4, 21], while some protocols work in small field [13]. We will evaluate MPC's actual performance in Sect. 8.4.4.

A zero-knowledge proof is a cryptographic protocol that allows one party, called the *prover*, to prove to another party, called the *verifier*, that a statement is true, without revealing any information other than the veracity of the statement. A ZKP must satisfy the following properties:

- **Completeness**: If the statement is true, the honest verifier should be convinced of this fact by an honest prover.
- **Soundness**: If, on the other hand, the statement is false, no cheating/malicious prover can convince the honest verifier to believe that it is true (except maybe with some small probability).
- **Zero-knowledge**: If the statement is true, no cheating/malicious verifier should be able to learn anything other than this fact. This is often proven using a simulation paradigm: for any cheating verifier, there exists some simulator that, given only the statement, can produce a transcript that is indistinguishable from an interaction between the honest prover and the cheating verifier.

ZKP was first conceived by Goldwasser et al. [45] in the 1980s. Since its invention, ZKP has found numerous applications in situations where honest behavior and privacy must be enforced at the same time (e.g., MPC [43]). It is only recently that the *privacy-preserving data mining* (PPDM) community, realizing the importance of enforcing correct behavior in data analysis, started adopting ZKPs in their solutions.

Noticeably, there is a line of work [11,29–32] that attempt to adapt or compose basic ZKPs into efficient and scalable primitives that support more practical verifications, required in large-scale analysis (e.g., a vector's L2-norm is bounded [29–31]). They are based on the finite field ZKPs of [19] and use random projection to reduce the number of large field arithmetic operations so that the techniques can be used on large-scale problems [29–32].

8.2.2 Privacy-Preserving Data Mining

The field of privacy preserving data mining deals with situations where a number of players perform mining tasks over their joint data, while preserving the secrecy of their data against other players. Usually the data is either partitioned across several servers or fully distributed among clients. There are privacy solutions for a number of data mining tasks, including classification [26], clustering [73] and decision tree learning [55]. These schemes use either *randomization* [25, 37] or *cryptographic* techniques [26, 55, 73, 74, 76] to protect privacy. The first approach, besides sacrificing accuracy, has been shown to provide very little privacy in many cases [49]. The second approach, on the other hand, enjoys provable privacy that is guaranteed by the cryptographic primitives, in their own adversary models and up to the information that they deem publishable. However, most methods are not practical at large scale, nor can they deal with realistic adversary models, where some malicious players may actively cheat to disrupt the computation. To hide information, these methods make heavy use of public-key cryptosystems [26, 55, 73, 74, 76]. Although the schemes may have the same asymptotic cost as the standard algorithms, the constant factor imposed by the public key operations is prohibitive.

8.2.3 Statistical Database Privacy

A common problem with secure multiparty computation methods is that they do not examine the possible leakage caused by the release of the final results. For example, a privately evaluated sum of two numbers, each held by a player, allows one player to compute the other player's value. Such leakage can be modeled and prevented using results from research in statistical database privacy, which studies the problem of releasing statistical patterns of the data, while preserving the privacy of each individual record.

Statistical database privacy has been extensively studied since the 1970s. The early results were mixed, mostly due to the lack of a proper notion of privacy. For example, some methods (e.g., [14,52]) only consider full disclosure as compromise, which is apparently too weak by today's standard. Motivated by the notion of semantic security in asymmetric key encryption, a newly emerging line of work

[8, 24, 34–36, 64] strive to provide a rigid privacy definition and protections, as rigorous as those in cryptography. Current results state that query responses need to be perturbed by random noise with sufficient variance in order to maintain privacy. We will draw heavily on their results when we discuss the notions of privacy in Sect. 8.3.1.

8.3 Preliminaries

We say that an adversary is *passive*, or *semi-honest*, if he tries to compute additional information about other players' data but still follows the protocol. An *active*, or *malicious* adversary, on the other hand, can deviate arbitrarily from the protocol, including inputting bogus data, producing incorrect computation, and aborting the protocol prematurely. Our scheme is secure against a hybrid threat model that includes both passive and active adversaries. We introduce the model in Sect. 8.5.

8.3.1 Notions of Privacy

Privacy is an elusive concept; different individuals may have very different understandings and expectations of privacy. For this, we must establish a rigorous definition of privacy before we proceed to the proposed approach.

8.3.1.1 Early Notions

Over the years many privacy definitions have been proposed. The early ones are *syntactic* in that they specify certain conditions that the released data must satisfy, without considering their *semantic* implications. Sweeney proposed the famous notion of k-anonymity [68, 69, 71]. Informally, k-anonymity requires that every record in the released dataset is syntactically indistinguishable among at least k records on the, so-called, *quasi-identifying attributes*, such as ZIP code, gender and date of birth. This can be achieved by syntactic *generalization* and *suppression* of values in these attributes. k-anonymity has been shown to be weak and vulnerable to various attacks [41, 58, 75, 78]. Moreover, it does not hide the presence of an individual within the database [63], nor his attributes [54, 58]. Subsequent variants, such as l-diversity [33, 58] and t-closeness [54], suffer from similar flaws [33].

The problem with syntactic definitions is that they do not target directly at a clear goal but focus on properties that "seem" right. These "intuitive" definitions fail to capture the essence of privacy. This motivates a very careful and formal treatment of *semantic* notion. In 1977 Dalenius offered a semantic notion of privacy breach for statistical databases: "If the release of the statistic S makes it possible to determine the (microdata) value more accurately than without access to S, a disclosure has taken place" [20]. However, this is impossible to prevent if a database is to provide some degree of utility [33]. This is because Dalenius's definition is based on a prior/posterior approach: the difference between the adversary's prior and posterior views about an individual (i.e., before and after accessing the database) should be small. This goal is in direct conflict with the purposes of using the data: after all, one expects to learn some non-trivial facts from the data. The facts could be laws of nature (e.g., smoking causes lung cancer) or global statistics (e.g., average height) of a population. Both may allow an adversary to infer an individual's information.

8.3.1.2 Differential Privacy

Dwork [33] proposed the notion of *differential privacy* that is an achievable semantic notion based on participation: "we move from comparing an adversary's prior and posterior views of an individual, to comparing the risk to an individual when included in, versus when not included in, the database." Differential privacy strikes a good balance between secrecy and utility. Formally it is defined as:

Definition 8.1 (Differential Privacy [35, 36]). $\forall \epsilon, \delta \geq 0$, an algorithm $\mathscr{A}$ gives (ϵ, δ)-differential privacy if for all $S \subseteq Range(\mathscr{A})$, for all data sets D, D' such that D and D' differ by a single record

$$\Pr[\mathscr{A}(D) \in S] \leq \exp(\epsilon) \Pr[\mathscr{A}(D') \in S] + \delta$$

$\mathscr{A}^f$ is said to be (ϵ, δ)-private if it gives (ϵ, δ)-differential privacy.

Differential privacy captures the intuition that the function is private if the risk to one's privacy does not substantially increase as a result of participating in the data set. It has been widely adopted by many recent works, such as [9,10,61,62,64] and has became the "gold standard" of privacy. There are several solutions achieving differential privacy for some machine learning and data mining algorithms (e.g., [1,9,10,35,57,60–62,64]). Most works require a trusted server hosting the entire data set and ignore the cases where the data sources may be malicious. Beimel et al. [5] proposes a distributed and differentially private scheme for binary sum functions, but it is also only secure in a semi-honest model.

Recently, there are some critics on differential privacy. Kifer et al. [51] point out that "participating in the database" is not equivalent to "appearing as a tuple in the database". For example, a user joining a social network may cause multiple edges to be added in the graph, thus affecting multiple tuples. Whether differential privacy, whose definition measures the distance between two database instances by the number of the tuples they differ (see Definition 8.1), protects this participation statues, depends on the data generation process. Some (e.g., independent) assumptions are necessary in order for differential privacy to provide adequate protection. Cormode [18] shows that a differentially private mechanism may allow an adversary to infer information about an individual with non-trivial accuracy. For example, we can learn a classifier (e.g., a Naive Bayes classifier) from the differentially private database responses and use that to predict that a smoking person, not necessarily included in the database, has a higher chance of getting lung cancer. In other words, if the database teaches us some patterns about the population, simply knowing that a sample is drawn from the same distribution allows one to predict its sensitive properties.

On this we take a practical perspective: for any data analysis to be useful, certain information must be learned from the data. It is the application's decision whether certain assumptions have to be made about the data generation process, as required by Kifer and Machanavajjhala [51], or some individuals must tolerate certain degree of privacy breach in order to obtain something of bigger value (e.g., insights about the causes of lung cancer). In the former case, the requirements in [51] are quite mild and fit many real-world applications. In the latter, privacy becomes more of a social issue (it is never a purely technical one) and should be handled accordingly. We believe that differential privacy is still a sound definition and we adopt it in this chapter.

For further reading, we refer the readers to [33] for an excellent survey on privacy notions and an in-depth discussion on the motivation of differential privacy.

8.4 Design Considerations

Our design was motivated by careful evaluation of goals, available resources and alternative solutions. In what follows, we elaborate on each of these aspects.

8.4.1 Design Goals

Our goal is to provide practical privacy solutions for real-world applications. To this end, we identify three properties that are essential to a practical privacy solution:

1. **Provable Privacy**: Its privacy must be rigorously proven against well formulated privacy definitions.
2. **Efficiency and Scalability**: It must have adequate efficiency at reasonably large scale, which is an absolute necessity for many of today's data mining applications. To support real-world applications, both the number of users and the number of data items per user are assumed to be in millions.
3. **Robustness**: It must be secure against realistic adversaries. Many computations either involve the participation of users, or collect data from them. Cheating of a small number of users is a realistic threat that the system must handle.

In what follows, we show that trivial composition of existing solutions cannot attain all three. Large-scale data analysis calls for a new privacy paradigm.

8.4.2 Issue of Field Sizes

Throughout the rest of the chapter, we will mention the issue of "field or integer size" several times. Unless otherwise noted, this refers to the bit length of the integers the system manipulates. Typical public key cryptosystems work in the "large" field (e.g., 1,000-bit for ELGamal or 160-bit for ECC) for security, while most programming languages support built-in small or regular fields (32 or 64-bit integers). We will show with concrete numbers that the costs for manipulating integers in different fields differ dramatically, resulting in huge performance improvement if the majority of the computation is done in small field.

8.4.3 Available Resources

During the past few years the landscape of large-scale distributed computing has changed dramatically. Many new resources and paradigms are nowadays available at very low cost and many computations that were infeasible at large scale in the past, are now running routinely. One notable trend is the rapid growth of "cloud computing", which refers to the model where clients purchase computing cycles and/or storage from a third-party provider over the Internet. Vendors are sharing their infrastructures and allowing general users access to gigantic computing capability. Industrial giants, such as Microsoft, Yahoo! and Google, are all key players in the game. Some cloud services (e.g., Amazon's Elastic Compute Cloud) are already available to the general public at very cheap prices.

The growth of cloud computing symbolizes the increased availability of large-scale computing power. We believe it is time to re-think the issue of privacy preserving data mining in light of such changes. There are several significant differences:

1. Cloud computing providers have very different incentives. Unlike traditional e-commerce vendors, who are naturally interested in users' data (e.g., purchases), the cloud computing providers's commodity (CPU cycles and disk space) is *orthogonal* to users' computation. Providers do not benefit directly from knowing the data or computation results, other than ensuring that they are correct.
2. The traditional image of client-server paradigm has changed. In particular, the users have much more control over the data and the computation. In fact, in many cases the cloud servers will be running code written by the customers. This is to be contrasted with traditional e-commerce, where there is a tremendous power imbalance between the service provider, who possesses all the information and controls what computation to perform, and the client users.
3. The servers are now clusters of hundreds or even thousands of machines, capable of handling huge amounts of data. They are not bottlenecks anymore.

Discrepancy of incentives and power imbalance have been identified as two major obstacles for the adoption of privacy technology by researchers examining privacy issues from legal and economic perspectives [2, 38]. Interestingly, both are greatly mitigated with the dawn of cloud computing. While traditional e-commerce vendors are reluctant to adopt privacy technologies, which may hinder their ability to benefit from analyzing user data and adds additional cost to their operation, cloud providers would happily comply with customers instructions regarding what computation to perform. And once a treasure for the traditional e-commerce vendors, user data is now almost a burden for the cloud computing providers: storing the data not only costs disk space, but also may entail certain liability, such as hosting illegal information. Some cloud providers may even choose not to store the data. For example, with Amazon's EC2 service, user data only persists during the computation.

We believe that cloud computing offers an extremely valuable opportunity for developing a new paradigm of practical privacy-preserving distributed computation: the existence of highly available, highly reputable, legally bounded service providers also provides a very important source of security. By tapping into this resource, we can build a heterogeneous system that can have privacy, scalability and robustness, all at once.

8.4.4   The Alternatives

In this subsection we discuss some alternative ways for building privacy-preserving applications. They can be classified into two models: *fully distributed* and *server-based*. The former model is represented by a large amount of work in the area of secure multiparty computation in cryptography. The latter includes mostly homomorphic encryption-based schemes, such as [11, 28, 76].

8.4.4.1   Generic Secure Multiparty Computation

In Sect. 8.2 we already surveyed the development in MPC and summarized their results. From the practitioners' perspective, these generic MPC protocols are mostly of theoretical interest. Reducing asymptotic complexity does not automatically make the schemes practical. These schemes tend to be complex, which imposes a huge barrier for developers not familiar with this area. Trying to support generic computation, most of them compile an algorithm into a (boolean or arithmetic) circuit. Not only the depth of such a circuit can be huge for complex algorithms, it is also very difficult, if not entirely impossible, to incorporate existing infrastructures and tools (e.g., ARPACK, LAPACK, MapReduce, etc.) into such computation. These tools are an indispensable part of our daily computing life and

Table 8.1 Performance comparison of existing MPC implementations

System	Adversary model	Benchmark	Run time (s)
Fairplay [59]	Semi-honest	Billionaires	1.25
FairplayMP [6]	Semi-honest	Binary tree circuit (512 Gates)	6.25
PSSW [67]	Semi-honest	AES encryption of 128-bit block	7
LPS [56]	Malicious	16-bit integer comparison	135

symbolize the work of many talents over many years. Re-building production-ready implementations is costly and error-prone, and generally not an option for most companies in our fast-pacing modern world.

Recently, there are several systems that implemented some of the MPC protocols. While this reflects a plausible attempt to bridge the gap between theory and practice, unfortunately, performance-wise none of the systems came close to providing satisfactory solutions for most large-scale real-world applications. Table 8.1 shows some representative benchmarks obtained by these implementations. We only show their performance in their most favorable model (i.e., semi-honest model), unless the whole system is built for the general setting. Using FairplayMP [6] as an example, adding two 64-bit integers is compiled into a circuit of 628 gates and 756 wires using its SFDL compiler. According to the benchmark in [6], evaluating such a circuit between two players takes about 7 s. With this performance, adding 10^6 vectors of dimensionality 10^6 each, which constitutes one iteration in our framework, takes 7×10^{12} s, or about 221,969 years!

8.4.4.2 Server-Based Elliptic Curve Cryptography Scheme

It has been shown that the conventional client-server paradigm can be augmented with homomorphic encryption to perform some computations with privacy (e.g., [11, 28, 76]). Still, such schemes are only marginally feasible for small to medium-scale problems due to the need to perform at least a linear number of large-field operations, even in the purely semi-honest model, as demonstrated by our benchmark in Sect. 8.8.2. Elliptic curve cryptography (ECC) schemes can mitigate the problem as ECC can reduce the size of the cryptographic field (e.g., a 160-bit ECC key provides the same level of security as a 1,024-bit RSA key). ECC cryptosystems, such as [65], are $(+, +)$-homomorphic which is ideal for private computation. However, ECC point addition requires one field inversion and several field multiplications. The operation is still orders of magnitude slower than adding 64 or 32-bit integers directly. According to our benchmark, inversion and multiplication in a 160-bit field take 0.0224 and 0.001 ms, respectively. Adding one million 10^6-element vectors takes about 260 days.

8.5 Principles and Architecture

In this section, we discuss the main features of our approach and elaborate on the considered threat model. Next, we present the proposed framework, called *Peers for Privacy*, in Sect. 8.6.

8.5.1 Principles

The lesson that we learn from the previous section is that, for large-scale problems, privacy and security must be added with low cost. In particular, those steps that dominate the computation should *not* be burdened with public-key cryptographic operations (even those "efficient" ones, such as ECC), and ideally should be carried out with the same cost as a direct, non-private implementation, simply because they have to be performed so many times. This is the major principle that guides our design. Specifically, our approach features the following:

- We present a general computation paradigm that decomposes an algorithm into a series of vector addition steps. Local computation (i.e., those performed on each client) can be arbitrary but global computation (i.e., those over all users data) is restricted to additions.
- The main computation is performed in *small* (e.g., 32 or 64-bit) field using privacy schemes, such as *verifiably secret sharing* (VSS).
- We use random projection and similar techniques to devise efficient ZK verification, with $O(\log m)$ or less cost, to enforce necessary security properties during the computation.

A few remarks are in order. We choose addition as our basic computation primitive for the following reasons. First, addition is a fundamental operation for aggregating information across samples. Many important statistics (e.g., expectation, moments, etc.) are global summations of local variables. Many statistical principles, such as maximum likelihood, are also based on addition. For example, under the i.i.d. assumption, log likelihood is the summation of local quantities across all samples. This simple addition-only primitive is therefore a surprisingly powerful model, supporting a large number of popular data mining and machine learning algorithms. Examples include linear regression, Naive Bayes, PCA, k-means, ID3 and EM, as has been demonstrated by numerous previous works, such as [11, 12, 15, 22, 28]. Moreover, it has been shown that all algorithms in the statistical query model [50] can be expressed in this form.

Second, private addition in the VSS paradigm does not require interaction, thus has extremely efficient implementations. In fact it can be as efficient as the direct, non-private computation if done over small field. In other words, privacy can be added to addition steps with almost zero cost. Moreover, addition is extremely easy to parallelize so aggregating a large amount of numbers is straightforward.

When it comes to enforcing proper behavior and limiting malicious influence, we notice that, since data analysis and learning algorithms learn from statistics from a large number of samples, it is often not necessary to perform verification on each element of user data. Rather, it suffices to bound some form of aggregates, such as its L2-norm. This is also inline with many perturbation theory results. In Sect. 8.6.3.2, we will use a concrete example to illustrate the effectiveness of such an approach. These statistics, or some of their properties, can be estimated using various probabilistic techniques, such as random projection. The benefit of such an approach is that it greatly reduces the number of large-field cryptographic operations and allows for practical ZK tools for verifying large-scale data. Our experiments show that, when the number of cryptographic operations are insignificant, even using the traditional ElGamal encryption (or commitment) with 1,024-bit key, the performance is adequate for large-scale problems. We summarize available random projection-based ZK tools in Sect. 8.6.3.1.

8.5.2 Architecture

Our proposed framework is called *Peers for Privacy*, or P4P. It was first introduced in [32]. The name comes from the feature that, during the computation, certain *aggregate* information is released. This is a very important technique that allows the private protocol to have high efficiency. It can be shown [32] that publishing such aggregate information does not harm privacy: individual traits are masked out in the aggregates and releasing them is safe. In other words, peers data mutually protects each other within the aggregates.

Let $\kappa > 1$ be a small integer. We assume that there are κ servers belonging to *different* service providers (e.g., Amazon's EC2 service and Microsoft's Azure Services Platform.[2]) We define a *server* as all the computation units under the control of a single entity. It can be a cluster of thousands of machines so that it has the capability to support a large number of users. From data protection perspective, it suffices to view them as a single entity.

Threat Model *Let $\alpha \in [0, 0.5)$ be the upper bound on the fraction of the dishonest users in the system.[3] Our scheme is robust against a computationally bounded adversary, whose capability of corrupting parties is modeled as follows:*

1. *The adversary may actively corrupt at most $\lfloor \alpha n \rfloor$ users, where n is the number of users.*
2. *In addition to 1, we also allow the same adversary to passively corrupt $\kappa - 1$ servers.*

---

[2]http://www.microsoft.com/azure/default.mspx.

[3]Most statistical algorithms need to bound the amount of noise in the data to produce meaningful results. This means that the fraction of cheating users is usually below a much lower threshold (e.g., $\alpha < 20\%$).

This model was proposed in [30] and is a special case of the general adversary structure introduced in [40, 47, 48] in that some of the participants are actively corrupted, while some others are passively corrupted by the same adversary *at the same time*. The model does not satisfy the feasibility requirements of [40, 47, 48]. We avoid the impossibility, by considering addition only computation.

The proposed model models realistic threats in our target applications. In general, users are not trustworthy. Some may be incentivized to bias the computation, some may have their machines corrupted. So we model them as active adversaries and our protocol ensures that active cheating of a small number of users will not exert large influence on the computation. The servers, on the other hand, are selling CPU cycles and disk space, something that is not related to user's computation or data. Deviating from the protocol causes them potential penalty (e.g., loss of revenue for incorrect results) but little benefit. Their threat is therefore passive. Corrupted servers are allowed to share data with corrupted users.

Treating "large institutional" servers as semi-honest, non-colluding players has already been established by various previous works [55, 73, 74, 76]. The growth of cloud computing and the availability of many vendors make this model more and more realistic. However, in previous works, the servers are not only semi-honest, but also "trusted", in that some user data is exposed to at least one of the servers (vertical or horizontal partitioned database). Our model does not have this type of trust requirement, as each server only holds a random share of the user data. This further reduces the server's incentive to try to benefit from user data (e.g., reselling it), because the information it has are just random numbers without the other shares. A compromise requires the collusion of *all* servers, which is a much more difficult endeavor. An interesting and nice consequence of this arrangement is that the service provider is relieved of the liability of hosting and computing secret or illegal data. This could be one of the technical solutions that some (e.g., [23]) envisions cloud providers will have to seek.

8.6 The Peers for Privacy Framework

Let n be the number of users. Let ϕ be a small (e.g., 32 or 64-bit) prime. We write $\mathbb{Z}_\phi$ for the additive group of integers modulo ϕ. Let a_i be private user data for user i and I be public information. Both can be matrices of arbitrary dimensions with elements from arbitrary domains. Our scheme supports any iterative algorithms whose $(t+1)$-th update can be expressed as

$$I^{(t+1)} = f(\sum_{i=1}^{n} d_i^{(t)}, I^{(t)}),$$

where $d_i^{(t)} = g(a_i, I^{(t)}) \in \mathbb{Z}_\phi^m$ is an m-dimensional data vector for user i computed locally. Typical values for both m and n can range from thousands to millions. Both f and g are in general non-linear.

In the SVD example that we will present, $I^{(t)}$ is the vector returned by ARPACK, g is matrix-vector product and f is the internal computation performed by ARPACK. As shown earlier, this simple primitive could be the basis for a large number of practical algorithms. A secure instantiation of such primitive, provides practical solutions for securing a large number of applications.

8.6.1 Private Computation

In the following, we only describe the protocol for one iteration, since the entire algorithm is simply a sequential invocation of the same protocol. The superscript is thus dropped from the notation. For simplicity, we only describe the protocol for the case of $\kappa = 2$. It is straightforward to extend it to support $\kappa > 2$ servers (by substituting the $(2, 2)$-threshold secret sharing scheme with a (κ, κ) one). Using more servers strengthens the privacy protection but also incurs additional cost. We do not expect the scheme will be used with a large number of servers. Let S_1 and S_2 denote the two servers. Leaving out validity and consistency check, which will be illustrated using the SVD example, the basic computation is carried out as follows:

1. User i generates a uniformly random vector $u_i \in \mathbb{Z}_\phi^m$ and computes $v_i = d_i - u_i$ mod ϕ. She sends u_i to S_1 and v_i to S_2.
2. S_1 computes $\mu = \sum_{i=1}^n u_i \mod \phi$ and S_2 computes $v = \sum_{i=1}^n v_i \mod \phi$. Then, S_2 sends v to S_1.
3. S_1 updates I with $f((\mu + v) \mod \phi, I)$.

It is straightforward to verify that if both servers follow the protocol, then the final result is indeed the sum of the user data vectors mod ϕ. This result will be correct if every user's vector lies in the specified bounds for L2-norm, which is checked by the ZKP in [30].

8.6.2 Provable Privacy

Theorem 8.1. *P4P's computation protocol leaks no information beyond the intermediate and final aggregates, if no more than $\kappa - 1$ servers are corrupted.*

The proof follows easily from the fact that both the secret sharing scheme (for the computation) and the Pedersen commitment scheme [19, 66], used in the ZK protocols, are information-theoretic private, as the adversary's view of the protocol is uniformly random and contains no information about user data. We refer the readers to [42] for details and formal definition of information-theoretic privacy.

As for the leakage caused by the released sums, first, for SVD, and some other algorithms, we are able to show that the sums can be approximated from the final result so they do not leak more information (see Sect. 8.7.6). For general computation, we draw on the works on differential privacy. Duan [27] has shown that, by using well-established results from statistical database privacy [8, 24, 36], under certain conditions, releasing the vector sums still maintains differential privacy.

In some situations verifying the conditions of [27] privately is non-trivial but this difficulty is not essential in our scheme. There are well-established results that prove that differential privacy, as well as adequate accuracy, can be maintained as long as the sums are perturbed by independent noise with variance calibrated to the number of iterations and the sensitivity of the function [8, 24, 36]. In our settings, it is trivial to introduce noise into our framework – each server, which is semi-honest, can add the appropriate amount of noise to their partial sums after all the vectors from users are aggregated. Calibrating noise level is also easy: all one needs are the parameters ϵ, δ, the total number of queries (mT in our case where T is the number of iterations) and the sensitivity of the function f, which is summation in our case, defined as [36]:

$$S(f) = \max_{D, D'} \| f(D) - f(D') \|_1,$$

where D and D' are two data sets differing by a single record and $\| \cdot \|_1$ denotes the L1-norm of a vector. Cauchy's Inequality states that

$$\left(\sum_{i=1}^{m} x_i y_i\right)^2 \le \left(\sum_{i=1}^{m} x_i^2\right)\left(\sum_{i=1}^{m} y_i^2\right)$$

For a user vector $a = [a_1, \ldots, a_m]$, let $x_i = |a_i|$, $y_i = 1$, we have

$$\|a\|_1^2 = \left(\sum_{i=1}^{m} |a_i|\right)^2 \le \left(\sum_{i=1}^{m} a_i^2\right)m = \|a\|_2^2 m$$

Since our framework bounds the L2-norm of a user's vector to below L, this means the sensitivity of the entire computation is at most $\sqrt{m}TL$.

Note that the perturbation does not interfere with our ZK verification protocols in any way, as the latter is performed between each user and the servers on the *original* data. Whether noise is necessary or not is dependent on the algorithm. For simplicity, we will not describe the noise process in our protocol explicitly.

8.6.3 Enforcing Correct Behavior

In this section we present zero-knowledge tools that are supported by the P4P framework for verifying proper behavior during the computation, and summarize their key properties. Next, we discuss issues related to restricting the influence of malicious players from corrupting the computation.

8.6.3.1 Zero-Knowledge Tools

The VSS scheme that the P4P framework is built upon is $(+, +)$-homomorphic in that if each of the secret share holders adds her shares of two secrets together, she obtains a valid share of the sum of the two secrets [17]. The transformation does not require any interaction between players. This property, augmented with similar homomorphism of the commitment scheme, allows for easy construction of commitments to new values. As a result, the P4P framework admits extremely efficient zero-knowledge tools for verifying proper behavior during the computation. These tools provide a wealth of means for various applications to enforce correctness. In this section, we present available ZK tools. Some are standard cryptographic primitives, while others are more complex ones. Detailed construction and proofs can be found in [11,19,30,32,66]. We summarize only their key properties here. For concreteness, we describe those tools that are based on the discrete log assumptions. They can also rely on RSA.

Let p and q be two large primes (e.g., 1,024-bit), such that $q | p - 1$. Let $\mathbb{Z}_p^*$ denote the multiplicative group of integers modulo the prime p. We use G_q to denote the unique subgroup of $\mathbb{Z}_p^*$ of order q. Let g and h be two generators of G_q, such that $x = \log_g h$ is unknown to anyone. g and h are made public. The discrete logarithm problem is assumed to be hard in G_q, i.e., given $y = g^x \mod p$, it is computationally infeasible to recover x. A Pedersen commitment [66] to a message $s \in \mathbb{Z}_q$ is computed as $\mathscr{C}(s, r) = g^s h^r \mod p$, where $r \leftarrow_R \mathbb{Z}_q$ is randomly drawn from $\mathbb{Z}_q$. This commitment is *information-theoretic* private. One could reveal $\mathscr{C}(s, r)$ without leaking any information about s [66]. The scheme is *computationally binding* in that, given s and r, a computationally bounded adversary not knowing $x = \log_g h$ cannot produce an s' and t' such that $g^s h^r = g^{s'} h^{r'}$ mod p, otherwise the discrete logarithm problem can be solved.

- **Homomorphic commitment**: A homomorphic commitment to an integer a with randomness r is a commitment, written as $\mathscr{C}(a, r)$, that satisfies the homomorphic property: $\mathscr{C}(a, r)\mathscr{C}(b, s) = \mathscr{C}(a + b, r + s)$. Homomorphic commitments allow one to construct commitments to new secrets from existing ones, without accessing to the original values. The above-mentioned Pedersen's commitment [66] is such a scheme.
- **ZKP of knowledge**: A prover who knows a and r (i.e., who knows how to open $\mathscr{A} = \mathscr{C}(a, r)$) can demonstrate that it has this knowledge to a verifier who knows only the commitment $\mathscr{A}$. The proof reveals nothing about a or r.

- **ZKP for equivalence**: Let $\mathscr{A} = \mathscr{C}(a, r)$ and $\mathscr{B} = \mathscr{C}(a, s)$ be two commitments to the same value a. A prover who knows how to open $\mathscr{A}$ and $\mathscr{B}$ can demonstrate to a verifier in zero knowledge that they commit to the same value.
- **ZKP for product**: Let $\mathscr{A}$, $\mathscr{B}$ and $\mathscr{C}$ be commitments to a, b, c respectively, where $c = ab$. A prover who knows how to open $\mathscr{A}$, $\mathscr{B}$, $\mathscr{C}$ can prove in zero knowledge to a verifier who has only the commitments that the relationship $c = ab$ holds among the values they commit to. If, say, a is made public, this primitive can be used to prove that $\mathscr{C}$ encodes a number that is multiple of a.
- **ZKP for L2-norm boundedness**: Let $a \in \mathbb{Z}_\phi^m$. A prover knowing a can produce a ZKP that proves that $\|a\|_2 < L$ for a public constant L. Duan and Canny [30] provides such a protocol with $O(\log m)$ cost. We refer to it as the L2-norm ZKP.

The ZK tools that the P4P framework supports are not restricted to the above examples. And indeed different algorithms may have different correctness conditions. We anticipate that there could be more application-dependent ZK tools tailored to each applications specific needs. P4P's VSS-based addition-only computation paradigm makes it easy to develop such tools. As an example, in Sect. 8.7.3 we will present a consistency ZKP, which was originally introduced in [32], that the data users input during each iteration is consistent. Again the protocol uses random projection and needs only constant number of large field operations.

8.6.3.2 Restricting Malicious Influence

In data analysis, it is often necessary to restrict the influence of a small number of samples. In a privacy-preserving setting, it is also essential to prevent a few malicious players from corrupting the computation. Interestingly, both can be handled in a uniform way. In this section, using the L2-norm as an example, we demonstrate how bounding some aggregate information of a player's input, as advocated by P4P's principles, is effective for achieving both goals.

Bounding the L2-norm of a user's vector has been proposed by Duan and Canny [30] as a practical way to restrict the amount of malicious influence on the computation a cheating user could cause. Its effectiveness can be shown from several perspectives. Firstly, notice that the result of the computation depends on the sums of n vectors. To drive the sums away from correct positions by a large amount, a malicious user must input a vector with sufficient "length", which is naturally measured by its L2-norm. This is especially evident for algorithms whose results are simply the vector sums (e.g., k-means). In this case even the precision of the final result is often measured by the L2-norm of the error vector (see e.g., Blum et al. [8]), which, by triangle inequality, is bounded by the sum of the L2-norms of all noise vectors. Duan et al. [31] provide a detailed discussion on the effectiveness of such verification in the computation of the k-means algorithm.

Secondly, many perturbation theories measure the perturbation to the system in terms of various forms of (matrix and vector) norms, many of which can be easily transformed into vector L2-norms. For example, let $\tilde{}$ denote the perturbed quantity and σ_i the i-th singular value of a matrix A. The classical Weyl and Mirsky theorems [70] bound the perturbation to A's singular values in terms of the spectral norm $\|\cdot\|_2$ and the Frobenius norm $\|\cdot\|_F$ of $E := A - \tilde{A}$, respectively:

$$\max_i |\tilde{\sigma}_i - \sigma_i| \le \|E\|_2 \text{ and } \sqrt{\sum_i (\tilde{\sigma}_i - \sigma_i)^2} \le \|E\|_F$$

The spectral norm can be bounded from above by Frobenius norm: $\|E\|_2 \le \|E\|_F$. And if each row, denoted a_i, of the matrix A is held by a user, the Frobenius norm of the matrix E can be expressed in terms of vector L2-norms:

$$\|E\|_F = \sqrt{\sum_{i=1}^{n} \|\tilde{a}_i - a_i\|_2^2}$$

Clearly bounding the vector L2-norm provides an effective way to bound the perturbation of the results. Similar techniques were also used in e.g., [28].

Finally, bounding the L2-norm can also be the basis of other, more specific checks. For instance, in a voting application, the protocol can be used with $L = 2$ and some simple tricks, which were introduced in [31], in order to ensure that each user only exercises one vote.

8.7 Private Large-Scale SVD

In the following, we use a concrete example, a private SVD scheme, to demonstrate how the P4P framework can be used to support private computation of popular algorithms.

8.7.1 Basics

Recall that for a matrix $A \in \mathbb{R}^{n \times m}$, there exists a factorization of the form

$$A = U\Sigma V^T, \tag{8.1}$$

where U and V are $n \times n$ and $m \times m$, respectively, and both have orthonormal columns. Σ is $n \times m$ with nonnegative real numbers on the diagonal sorted in descending order and zeros off the diagonal. Such a factorization is called a *singular value decomposition* of A. The diagonal entries of Σ are called the *singular values* of A. The columns of U and V are left (resp. right) *singular vectors* for the corresponding singular values.

SVD is a very powerful technique that forms the core of many data mining and machine learning algorithms. Let $r = rank(A)$ and u_i, v_i be the column vectors of U and V, respectively. Equation 8.1 can be rewritten as $A = U \Sigma V^T = \sum_{i=1}^{r} \sigma_i u_i v_i^T$, where σ_i is the ith singular value of A. Let $k \leq r$ be an integer parameter. We can approximate A by $A_k = U_k \Sigma_k V_k^T = \sum_{i=1}^{k} \sigma_i u_i v_i^T$. It is known that of all rank-k approximations, A_k is optimal in the Frobenius norm sense. The k columns of U_k (resp. V_k) give the optimal k-dimensional approximation to the columnspace (resp. rowspace) of A. This dimensionality reduction preserves the structure of original data, while considering only essential components of the matrix. It usually filters out noise and improves the performance of data mining tasks.

We use a popular eigensolver, ARPACK [53] (ARnoldi PACKage), and its parallel version PARPACK. ARPACK consists of a collection of Fortran77 subroutines for solving large-scale eigenvalue problems. The package implements the Implicitly Restarted Arnoldi Method (IRAM) and allows one to compute a few, say k, eigenvalues/eigenvectors with user specified features, such as those of largest magnitude. Its storage complexity is $nO(k) + O(k^2)$, where n is the size of the matrix. ARPACK is a freely-available, yet powerful tool. It is best suited for applications whose matrices are either sparse or not explicitly available: it only requires the user code to perform some "action" on a vector, supplied by the solver, at every IRAM iteration. This action is simply matrix-vector product in our case. Such a reverse communication interface works seamlessly with P4P's aggregation protocol.

8.7.2 Private SVD Scheme

In our setting the rows of A are distributed across all users. We use $A_{i*} \in \mathbb{R}^m$ to denote the m-dimensional row vector owned by user i. From Eq. 8.1, and the fact that both U and V are orthonormal, it is clear that $A^T A = V \Sigma^2 V^T$, which implies that $A^T A V = V \Sigma^2$. A straightforward way is then to compute $A^T A = \sum_{i=1}^{n} A_{i*}^T A_{i*}$ and solve for the eigenpairs of $A^T A$. The aggregate can be computed using our private vector addition framework. This is a distributed version of the method proposed in [8] and does not require the consistency protocol that we will introduce later. Unfortunately, this approach is not scalable as the cost for each user is $O(m^2)$.

Suppose that $m = 10^6$, and each element is a 64-bit integer. Then, $A_{i*}^T A_{i*}$ is 8×10^{12} bytes, or about 8 TB. The communication cost for each user is then 16 TB (she must send shares to two servers). This is a huge overhead, both communication- and computation-wise. Usually the data is very sparse and it is a common practice to reduce cost by utilizing the sparsity. Unfortunately, sparsity does not help in a privacy-respecting application: revealing which elements are non-zero is a huge privacy breach and the users are forced to use the dense format.

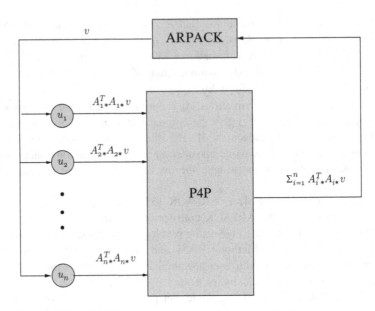

Fig. 8.1 Private SVD with P4P

We propose the following scheme which reduces the cost dramatically. We involve the users in the iteration and the total communication (and computation) cost per iteration is only $O(m)$ for each user. The number of iterations required ranges from tens to over a thousand. This translates to a maximum of a few GB data communicated for each user for the *entire* protocol, which is much more manageable.

One server, say S_1, will host an ARPACK engine and interact with its reverse communication interface. In our case, since $A^T A$ is symmetric, the server will use dsaupd, ARPACK's double precision routine for symmetric problems, and ask for the k largest (in magnitude) eigenvalues. At each iteration, dsaupd returns a vector v to the server code and asks for the matrix-vector product $A^T A v$. Notice that $A^T A v = \sum_{i=1}^{n} A_{i*}^T A_{i*} v$, each term in the summation is computable by each user locally in $O(m)$ time (by computing the inner product $A_{i*} \cdot v$ first) and the result is an m-vector. The vector can then be provided as input to the P4P computation, which aggregates the vectors across all users privately. The aggregate is the matrix-vector product, which can be returned to ARPACK for another iteration. This process is illustrated in Fig. 8.1.

The above method is known to have a sensitivity problem, i.e., a small perturbation to the input could cause a large error in the output. In particular, the error is $O(\|A\|^2/\sigma_k)$ [72]. Fortunately, most applications (e.g., PCA) only need the k largest singular values (and their singular vectors). This error is usually not a problem for those applications, since for the principal components $O(\|A\|^2/\sigma_k)$ is small. In fact, there is no noticeable inaccuracy in our test applications (i.e., latent semantic analysis for document retrieval). For general problems, the stable way is to compute the eigenpairs of the matrix

$$H = \begin{bmatrix} 0 & A^T \\ A & 0 \end{bmatrix}$$

It is straightforward to adopt our private vector addition framework to compute matrix-vector product with H. For simplicity, we will not elaborate on this.

8.7.3 Enforcing Data Consistency

During the iteration, user i should input $d_i = A_{i*}^T A_{i*} v$. However, a cheating user could input something completely different. This threat is different from inputting bogus data at the beginning and using it consistently throughout the iterations. The latter only introduces noise to the computation but generally does not affect the convergence. The L2-norm ZKP is effective in bounding the noise but does not help in enforcing consistency. The former, on the other hand, may cause the computation not to converge at all. This generally is a problem for iterative algorithms and is more than simply testing the equality of vectors: the task is complicated by the local function that each user uses to evaluate on her data, i.e., she is not simply inputting her private data vector, but some (possibly non-linear) function of it. In the case of SVD, the system needs to ensure that user i uses the same A_{i*} (to compute $d_i = A_{i*}^T A_{i*} v$) in all the iterations, not that she inputs the same vector.

Duan et al. [32] introduced a novel zero-knowledge tool that ensures that the correct data is used. The protocol is probabilistic and relies on random projection. That is, the user is asked to project her original vector and her result of the current round onto some random direction. It then tests the relation of the two projections. The protocol, which is listed below, catches cheating with high probability but only involves very few expensive large field operations.

8.7.3.1 Consistency Check Protocol

Since the protocol is identical for all users, we drop the user subscript for the rest of the chapter. Let $a \in \mathbb{Z}_\phi^m$ be a user's original vector (i.e., her row in the matrix A). The correct user input to this round should be $d = a^T a v$. For two vectors x and y, we use $x \cdot y$ to denote their inner product.

1. After the user inputs her vector d, in the form of two random vectors $d^{(1)}$ and $d^{(2)}$ in $\mathbb{Z}_\phi^m$, one to each server, such that $d = d^{(1)} + d^{(2)} \mod \phi$, S_1 broadcasts a random number r. Using r as the seed and a public PRG (pseudo-random generator), all players generate a random vector $c \in_R \mathbb{Z}_\phi^m$.

2. For $j \in \{1,2\}$, the user computes $x^{(j)} = c \cdot a^{(j)} \mod \phi$, $y^{(j)} = a^{(j)} \cdot v$ $\mod \phi$. Let $x = x^{(1)} + x^{(2)}$, $y = y^{(1)} + y^{(2)}$, $z = xy$. Let $w = (c \cdot a)(a \cdot v) - xy$. The user commits $\mathscr{X}^{(j)}$ to $x^{(j)}$, $\mathscr{Y}^{(j)}$ to $y^{(j)}$, $\mathscr{Z}$ to z, and $\mathscr{W}$ to w. She also constructs two ZKPs: (1) $\mathscr{W}$ encodes a number that is a multiple of ϕ. (2) $\mathscr{Z}$ encodes a number that is the product of the two numbers encoded in $\mathscr{X}$ and $\mathscr{Y}$, where $\mathscr{X} = \mathscr{X}^{(1)} \mathscr{X}^{(2)}$ and $\mathscr{Y} = \mathscr{Y}^{(1)} \mathscr{Y}^{(2)}$. She sends all commitments and ZKPs to both servers.

3. The user opens $\mathscr{X}^{(j)}$ and $\mathscr{Y}^{(j)}$ to S_j, who verifies that both are computed correctly. Both servers verify the ZKPs. If any of them fails, the user is marked as FAIL and the servers terminate the protocol with her.

4. For $j \in \{1,2\}$, the user computes $\tilde{z}^{(j)} = c \cdot d^{(j)} \mod \phi$, $\tilde{z} = \tilde{z}^{(1)} + \tilde{z}^{(2)}$ and $\tilde{w} = c \cdot d - \tilde{z}$. She commits $\tilde{\mathscr{Z}}^{(1)}$ to $\tilde{z}^{(1)}$, $\tilde{\mathscr{Z}}^{(2)}$ to $\tilde{z}^{(2)}$, and $\tilde{\mathscr{W}}$ to $\tilde{w}$. She constructs the following two ZKPs: (1) $\tilde{\mathscr{W}}$ encodes a number that is a multiple of ϕ and (2) $\tilde{\mathscr{Z}} \tilde{\mathscr{W}}$ and $\mathscr{Z} \mathscr{W}$ encode the same value. She sends all the commitments and ZKPs to both servers.

5. The user opens $\tilde{\mathscr{Z}}^{(j)}$ to S_j, who verifies that it is computed correctly. Both servers verify the two ZKPs. They mark the user as FAIL if any of the verifications fails and terminate the protocol with her.

6. Both servers output PASS.

8.7.3.2   Issue of Group Sizes

There are two large primes, p and q, and four groups involved in the protocol: (i) the large, multiplicative group $\mathbb{Z}_p^*$, used for commitments, (ii) the additive group of its exponents $\mathbb{Z}_q$ (which is also large), where the committed pre-image values and the random numbers are in, (iii) the "small" group $\mathbb{Z}_\phi$, used for additive secret-sharing, and (iv) the group of all integers. All the commitments, such as $\mathscr{X}^{(j)}$ and $\mathscr{Y}^{(j)}$, are computed in $\mathbb{Z}_p^*$, so standard cryptographic assumptions still apply. The inputs to the commitments, which can be user's data or some intermediate results, are either in $\mathbb{Z}_\phi$ or in the integer group (without bounding their values).

Restricting commitment inputs to small group does not compromise the security of the scheme, since the outputs are still in the large group. Using Pederson's commitment as an example, the hiding property is guaranteed by the random numbers that are generated in the *large* group of $\mathbb{Z}_q$ for each commitment. Breaking the binding property is equivalent to solving the discrete logarithm problem in $\mathbb{Z}_p^*$ [66].

The protocol makes it explicit which group a number is in, using the   mod operator (i.e., $x = g(y) \mod \phi$ restricts x to be in $\mathbb{Z}_\phi$, while $x = g(y)$ means x can be in the whole integer range). The protocol assumes that $p, q \gg \phi$. This ensures that the numbers that are in the integer group (x, y, z, w in step 2 and $\tilde{z}, \tilde{w}$ in step 4) are much less than p to avoid modular reduction when their commitments are produced. This is true for most realistic deployments, since ϕ is typically 64 bits or less, while both p and q are 1,024 bits or more.

Theorem 8.2 proves that the transition from $\mathbb{Z}_\phi$ to integer fields or $\mathbb{Z}_q$ only causes the protocol to fail with extremely low probability:

Theorem 8.2. *Let O be the output of the Consistency Check protocol. Then*

$$\Pr(O = PASS | d = a^T a v) = 1$$

and

$$\Pr(O = PASS | d \neq a^T a v) \leq \frac{1}{\phi}$$

Furthermore, the protocol is zero-knowledge.

Proof. If computed correctly, both w and $\tilde{w}$ are multiples of ϕ due to modular reduction. Because of homomorphism, the equivalence ZKP that $\tilde{\mathscr{Z}}\tilde{\mathscr{W}}$ and $\mathscr{Z}\mathscr{W}$ encode the same value is to verify that $c \cdot d = c \cdot (a^T a v)$.

Completeness: If the user performs the computation correctly, she should input $d = a^T a v$ into this round of computation. All the verifications should pass. The protocol outputs PASS with probability 1.

Soundness: Suppose that $d \neq a a^T v$. The user is forced to compute the commitments $\mathscr{X}^{(1)}, \mathscr{X}^{(2)}, \mathscr{Y}^{(1)}, \mathscr{Y}^{(2)}$, and $\tilde{\mathscr{Z}}^{(1)}, \tilde{\mathscr{Z}}^{(2)}$ faithfully since she has to open them to at least one of the servers. The product ZKP at step 2 forces the number encoded in $\mathscr{Z}$ to be xy, which differs from $c \cdot (a^T a v)$ by w. Due to homomorphism, at step 4, $\tilde{\mathscr{Z}}$ encodes a number that differs from $c \cdot d$ by $\tilde{w}$. The user could cheat by lying about w or $\tilde{w}$, i.e., she could encode some other values in $\mathscr{W}$ and $\tilde{\mathscr{W}}$ to adjust for the difference between $c \cdot d$ and $c \cdot (a^T a v)$, hoping to pass the equivalence ZKP. However, assuming the soundness of the ZKPs used, the protocol forces both to be multiple of ϕ (steps 2 and 4), so she could succeed only when the difference between $c \cdot d$, which she actually inputs to this round, and $c \cdot (a^T a v)$, which she should input, is some multiple of ϕ. Since c is made known to her *after* she inputs d, the two numbers are totally unpredictable and random to her. The probability that $c \cdot d - c \cdot (a^T a v)$ is a multiple of ϕ is only $1/\phi$, which is the probability of her success.

Finally, the protocol consists of a sequential invocation of some well-established zero-knowledge proofs. By the sequential composition theorem of [44], the whole protocol is also zero-knowledge.

As a side note, all the ZKPs can be made non-interactive using the Fiat-Shamir paradigm [39]. It is also much more light-weight than the L2-norm ZKP [30]: the number of large field operations is *constant*, as opposed to $O(\log m)$ in the L2-norm ZKP. The private SVD computation thus involves only one L2-norm ZKP at first round, and one light verification for each of the subsequent rounds.

8.7.4 Dealing with Real Numbers

In their simplest forms, the cryptographic tools only support computation over integers. In most domains, however, applications typically have to handle real numbers. In the case of SVD, even if the original input matrix contains only integer entries, it is likely that real numbers appear in the intermediate (e.g., the vectors returned by ARPACK) and the final results. Because of the linearity of the P4P computation, we can use a simple linear digitization scheme to convert between real numbers in the application domain and Z_ϕ, P4P's integer field.

Let $R > 0$ be the bound of the maximum absolute value application data can take. The integer field provides $|\phi|$ bits resolution. This means the maximum quantization error for one variable is $R/\phi = 2^{|R|-|\phi|}$. Summing across all n users, the worst case absolute error is bounded by $n2^{|R|-|\phi|}$. In practice $|\phi|$ can be 64, and $|R|$ can be around e.g., 20 (this gives a range of $[-2^{20}, 2^{20}]$). With $n = 10^6$, this gives a maximum absolute error of under 1 over a million.

8.7.5 Complete SVD Protocol

Let Q be the set of qualified users initialized to the set of all users. The entire private SVD method is summarized as follows:

1. **Input:** The user first provides an L2-norm ZKP [30] on a with a bound L, i.e., she submits a ZKP that $\|a\|_2 < L$. This step also forces the user to commit to the vector a. Specifically, at the end of this step, S_1 and S_2 have $a^{(1)} \in Z_\phi$ and $a^{(2)} \in Z_\phi$, respectively, such that $a = a^{(1)} + a^{(2)} \mod \phi$. Users who fail this ZKP are excluded from subsequent computation.

2. Repeat the following steps until the ARPACK routine indicates convergence or stops after certain number of iterations:

   (a) **Consistency Check:** When dsaupd returns control to S_1 with a vector, the server converts the vector to $v \in Z_\phi^m$ and sends it to all users. The servers execute the consistency check protocol for each user.

   (b) **Aggregate:** For any users who are marked as FAIL, or fail to respond, the servers simply ignore their data and exclude them from subsequent computation. Q is updated accordingly. For this round, they compute $s = \sum_{i \in Q} d_i$ and S_1 returns it as the matrix-vector product to dsaupd, which executes another iteration.

3. **Output:** S_1 outputs

$$\Sigma_k = diag(\sigma_1, \sigma_2, \ldots, \sigma_k) \in \mathbb{R}^{k \times k}$$
$$V_k = [v_1, v_2, \ldots, v_k] \in \mathbb{R}^{m \times k},$$

with $\sigma_i = \sqrt{\lambda_i}$, where λ_i is the ith eigenvalue and v_i the corresponding eigenvector computed by ARPACK, $i = 1, \ldots, k$, and $\lambda_1 \geq \lambda_2 \ldots \geq \lambda_k$.

Regarding the accuracy of the result produced by this protocol in the presence of *actively* cheating users, we have:

Theorem 8.3. *Let n_c be the number of cheating users. We use $\tilde{\cdot}$ to denote perturbed quantity and σ_i the i-th singular value of matrix A. Assuming that honest users vector L2-norms are uniformly random in $[0, L)$ and $n_c \ll n$, then*

$$\sqrt{\frac{\sum_i (\tilde{\sigma}_i - \sigma_i)^2}{\sum_i \sigma_i^2}} < 2\sqrt{\frac{n_c}{n}}$$

Proof. The classic Weyl and Mirsky theorems [70] bound the perturbation to A's singular values in terms of the Frobenius norm $\| \cdot \|_F$ of $E := A - \tilde{A}$:

$$\sqrt{\sum_i (\tilde{\sigma}_i - \sigma_i)^2} \leq \|E\|_F$$

In our case each row a_i of A is held by a user, we have $\|E\|_F = \sqrt{\sum_{i=1}^n \|\tilde{a}_i - a_i\|_2^2}$. Since the protocol ensures that $\|a_i\|_2 < L$ for all users, we have

$$\sqrt{\sum_i (\tilde{\sigma}_i - \sigma_i)^2} \leq \sqrt{\sum_{i=1}^n \|\tilde{a}_i - a_i\|_2^2} < \sqrt{n_c} L$$

Let $\xi = \sqrt{\sum_i (\tilde{\sigma}_i - \sigma_i)^2} / \sqrt{\sum_i \sigma_i^2}$, and assuming that honest users vector L2-norms are uniformly random in $[0, L)$ and $n_c \ll n$, then

$$\xi = \frac{\sqrt{\sum_i (\tilde{\sigma}_i - \sigma_i)^2}}{\|A\|_F} < \frac{\sqrt{n_c} L}{0.5\sqrt{(n - n_c)L}} \approx 2\sqrt{\frac{n_c}{n}}$$

The scheme is also quite robust against users' failures. During our tests, reported in Sect. 8.8, we simulated a fraction of random users "dropping out" of each iteration. Even when up to 50 % of the users dropped, for all our test sets, the computation still converged without noticeable loss of accuracy, measured by residual error (see Sect. 8.8.1) using the final matrix with failed users data ignored. This allows us to handle malicious users who actively try to disrupt the computation and those who fail to response due to technical problems (e.g., network failure) in a uniform way.

8.7.6 Privacy Analysis

Note that the protocol does not compute U_k. This is intentional. U_k contains information about user data: the ith row of U_k encodes user i's data in the k-dimensional subspace and should not be revealed at all in a privacy-respecting application. V_k, on the other hand, encodes "item" data in the k-dimensional subspace (e.g., if A is a user-by-movie rating matrix, the items will be movies). In most applications (e.g., [11]) the desired information can be computed from the singular values (Σ_k) and the right singular vectors (V_k^T). At each iteration, the protocol reveals the matrix-vector product $A^T A v$ for some vectors v. This is not a problem because the final results Σ_k and V_k^T already give an approximation of $A^T A$ ($A^T A = V \Sigma^2 V^T$). A simulator with the final results can approximate the intermediate sums. Therefore, the intermediate aggregates do not reveal more information.

8.8 Implementation and Evaluation

The P4P framework, including the SVD protocol, has been implemented in Java using JNI and a NativeBigInteger implementation from I2P.[4] We run several experiments. The server is a 2.50 GHz Xeon E5420 with 32 GB memory, the clients are 2.00 GHz Xeon E5405 with 800 MB memory allocated to the tests. In all the experiments, ϕ is set to be a 62-bit integer and p, q 1,024-bit.

We evaluated our implementation on three data sets: the Enron Email Data Set [16], EachMovie (EM) and a randomly generated dense matrix (RAND). The Enron corpus contains email data from 150 users, spanning a period of about 5 years (January 1998–December 2002). Our test was run on the social graph defined by the email communications. The graph is represented as a 150×150 matrix A, with $A(i, j)$ being the number of emails sent by user i to user j. EachMovie is a well-known test data set for collaborative filtering. It comprises ratings of 1,648 movies by 74,424 users. Each rating is a number in the range $[0, 1]$. Both the Enron and EachMovie data sets are very sparse, with densities 0.0736 and 0.0229, respectively. To test the performance of our protocol on dense matrices, we generated randomly a $2,000 \times 2,000$ matrix with entries chosen in the range $[-2^{20}, 2^{20}]$. The properties of the datasets are summarized in Table 8.2.

---

[4]http://www.i2p2.de/.

Table 8.2 Characteristics of the datasets

Dataset	Dimensions	Density	Range	Type
Enron	150×150	0.0736	$[0, 1{,}593]$	Social graph
EM	$74{,}424 \times 1{,}648$	0.0229	$[0, 1.0]$	Movie ratings
RAND	$2{,}000 \times 2{,}000$	1.0	$[-2^{20}, 2^{20}]$	Random

Table 8.3 Round complexity and precision

Enron	k	10	20	30	40	50	60	70	80	90	100
	N	67	97	122	162	109	137	172	167	171	169
	$\epsilon (\times 10^{-8})$	0.00049	0.0021	0.0046	0.0084	0.0158	0.0452	0.121	0.266	0.520	1.232
EM	k	10	20	30	40	50	60	70	80	90	100
	N	70	140	254	222	276	371	322	356	434	508
	$\epsilon (\times 10^{-12})$	0.470	0.902	1.160	1.272	1.526	1.649	1.687	2.027	2.124	2.254
RAND	k	10	20	30	40	50	60	70	80	90	100
	N	304	404	450	480	550	700	770	720	810	800
	$\epsilon (\times 10^{-9})$	3.996	3.996	3.996	3.996	3.996	3.996	3.996	3.996	3.996	3.996

8.8.1 Precision and Round Complexity

We measured two quantities: N, the number of IRAM iterations until ARPACK indicates convergence, and ϵ, the relative error. N is the number of matrix-vector computation that was required for the ARPACK to converge. It is also the number of times P4P aggregation is invoked. The error ϵ measures the maximum relative residual norm among all eigenpairs computed:

$$\epsilon = \max_{i=1,\dots,k} \frac{\| A^T A v_i - \lambda_i v_i \|_2}{\| v_i \|_2}$$

Table 8.3 summarizes the results. In all these tests, we used machine precision as the tolerance input to ARPACK. The accuracy we obtained is very good: ϵ remains very small for all tests (10^{-12}–10^{-8}). In terms of round complexity, N ranges from under 100 to a few hundreds. For comparison, we also measured the number of iterations required by ARPACK when we perform the matrix-vector multiplication directly without the P4P aggregation. In all experiments, we found no difference in N between this direct method and our private implementation.

8.8.2 Performance

We measured both running time and communication cost of the scheme. We focused on server load, since each user only needs to handle her own data so is not a bottleneck. We present the case with $\kappa = 2$ servers and measured the work on the server hosting the ARPACK engine since it shares the most load. Extending the analysis to $\kappa > 2$ servers is straightforward as the cost for the servers is linear in κ and the performance is still adequate for small κ.

The implementation confirmed our observations about the difference in costs for manipulating large and small integers. With 1,024-bit key length, one exponentiation within the multiplicative group $\mathbb{Z}_p^*$ takes 5.86 ms. Addition and multiplication of two numbers, also within the group, take 0.024 and 0.062 ms, respectively. In contrast, adding two 64-bit integers, which is the basic operations P4P framework performs, needs only 2.7×10^{-6} ms. The product ZKP takes 35.7 ms verifier time and 24.3 ms prover time. The equivalence ZKP takes no time since it is simply revealing the difference of the two random numbers used in the commitments [66]. For each consistency check, the user needs to compute nine commitments, three product ZKPs, one equivalence ZKP and four large integer multiplications. The total cost is 178.63 ms for each user. For every user, each server needs to spend 212.83 ms on verification.

For our test data sets, it takes 74.73 s of server time to validate and aggregate all 150 Enron users data on a *single* machine (each user needs to spend 726 ms to prepare the zero-knowledge proofs). This translates into a total of 5,000 s or 83 min spent on private P4P aggregation to compute $k = 10$ singular-pairs. To compute the same number of singular pairs for EachMovie, aggregating all users data takes about 6 h (again on a single machine) and the total time for 70 rounds is 420 h. Note that the total includes *both* verification and computation so it is the cost of a complete run. The server load appears large but actually is very inexpensive. The aggregation process is trivially parallelizable and using a cluster of, say 200 nodes, will reduce the running time to about 2 h. This amounts to a very insignificant cost for most service providers: Using Amazon EC2's price as a benchmark, it costs \$0.80 per hour for 20 EC2 Compute Units (8 virtual cores with 2.5 EC2 Compute Units each). Data transfer price is \$0.100 per GB. The total cost for computing SVD for a system with 74,424 users is merely about \$15, including data transfer and adjusted for difference in CPU performance between our experiments and EC2.

The communication overhead is also very small since the protocol passes very few large integers. The extra communication per client for one L2-norm ZKP is under 50 kb, and under 100 bytes for the consistency check. This is significantly smaller than the size of an average web page. The additional workload for the server is less than serving an extra page to each user.

To compare with alternative solutions, we implemented a method based on homomorphic encryption which is a popular private data mining technique (see e.g., [11,76]). We did not try other methods, such as the "add/subtract random" approach, with players adding their values to a running total, because they do not allow for verification of user data thus are insecure in our model. We tested both ElGamal and Paillier encryptions with the same security parameter as our P4P experiments (i.e., 1,024-bit key). With the homomorphic encryption approach, it is almost impossible to execute the ZK verification (although there is a protocol [11]) as it takes hours to verify one user. So we only compared the time needed for computing the aggregates. Figure 8.2 shows the ratios of running time between homomorphic encryption and

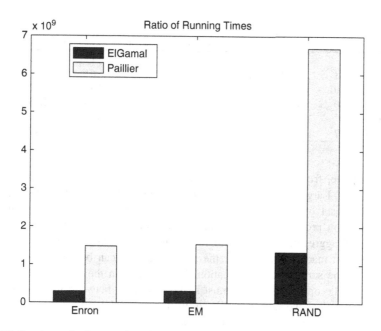

Fig. 8.2 Runtime ratios between homomorphic encryption based solutions and P4P

P4P for SVD on the three data sets. P4P is at least eight orders of magnitude faster in all cases for both ElGamal and Paillier. And this translates to tens of millions of dollars of cost for the homomorphic encryption schemes if the computation is done using Amazon's EC2 service not even counting data transfer expenses.

8.8.3   Scalability

We also experimented with a few very large matrices, with dimensionality ranging from tens of thousands to over a hundred million. They are document-term or user-query matrices that are used for latent semantic analysis. To facilitate the tests, we did not include the data verification ZKPs, as our previous benchmarks show they amount to an insignificant fraction of the cost. These results are meant to demonstrate the capability of our system, which we have shown to maintain privacy at very low cost, to handle large data sets at various configurations.

Table 8.4 summarizes some of the results. The running time measures the time of a complete run, i.e., from the start of the job till the results are safely written to disk. It includes both the computation time of the server (including the time spent on invoking the ARPACK engine) and the clients (which are running in parallel), and the communication time.

Table 8.4 SVD of large matrices

n	m	k	No. frontend processors	Time (h)	Iterations
100,443	176,573	200	32	1.4	1,287
12,046,488	440,208	200	128	6.0	354
149,519,201	478,967	250	128	8.3	1,579
37,389,030	366,881	300	128	9.1	1,839
1,363,716	2,611,186	200	1	14.8	1,260
33,193,487	1,949,789	200	128	28.0	1,470

In the table, frontend processors refer to the machines that interact with the users directly. Large-scale systems usually use multiple frontend machines, each serving a subset of the users. This is also a straightforward way to parallelize the aggregation process, i.e., each frontend machine receives data from a subset of users and aggregates them before forwarding to the server. On one hand, the more frontend machines the faster the sub-aggregates can be computed. On the other hand, the server's communication cost is linear in the number of frontend processors. The optimal solution must strike a balance between the two. Due to resource limitation, we were not able to use the optimal configuration for all our tests. The results are feasible even in these sub-optimal cases.

8.9   Conclusion and Future Trends

P4P's source code has been made available for general public.[5] The framework demonstrated that cryptographic building blocks can work harmoniously with existing tools, providing privacy without degrading their efficiency. As more and more data is being collected, large-scale data analysis has become a necessity in our daily life. It is predicted that in 2020, the sheer volume of digital information will reach 35 trillion gigabytes.[6] Be it scientific insights, or business intelligence, we learn invaluable information from the data. The social benefit provided by large-scale data analysis is huge, but so is the danger of privacy breach. A very urgent mission for the privacy community is to develop practical technologies that can keep pace with the rapidly growing need for large data. We believe P4P's paradigm provides an encouraging direction for this task. This belief is based on the following:

- P4P's pursuit for provably strong privacy guarantees adequate protection.
- P4P's main computation is based on VSS over small field, which, we emphasize again, has the same cost as non-private computation.

---

[5]http://bid.berkeley.edu/projects/p4p/.

[6]http://www.teradata.com/business-needs/Big-Data-Analytics/.

- P4P's ZK verification tools only involves a small number (constant or $O(\log m)$) large field cryptographic operations.

In other words, P4P allows privacy to be added with negligible cost. Furthermore, the VSS paradigm will become more realistic as more and more computing entities (e.g., cloud computing providers) are available. We expect that more privacy-preserving applications will be built upon this framework and principles.

While P4P provides practical tools for developers to build privacy-preserving real-world applications, we see it only as a promising start. Some applications, such as those that cannot be decomposed into addition steps, won't fit in the P4P model. Moreover, its reliance on multiple servers (for secret sharing) may also hinder its adoption in some cases. New privacy paradigms and techniques are required to meet the diverse needs for large-scale privacy preservation.

References

1. Alaggan, M., Gambs, S., Kermarrec, A.M.: Private similarity computation in distributed systems: from cryptography to differential privacy. In: Principles of Distributed Systems. Lecture Notes in Computer Science. Springer, Berlin/New York (2011)
2. Alderman, E., Kennedy, C.: The Right to Privacy. DIANE, Collingdale (1995)
3. Beaver, D., Goldwasser, S.: Multiparty computation with faulty majority. In: CRYPTO'89, Santa Barbara
4. Beerliová-Trubíniová, Z., Hirt, M.: Perfectly-secure mpc with linear communication complexity. In: TCC 2008, New York, pp. 213–230. Springer (2008)
5. Beimel, A., Nissim1, K., Omri, E.: Distributed private data analysis: simultaneously solving how and what. In: CRYPTO 2008, Santa Barbara (2008)
6. Ben-David, A., Nisan, N., Pinkas, B.: Fairplaymp: a system for secure multi-party computation. In: CCS'08, Alexandria, pp. 257–266 (2008)
7. Ben-Or, M., Goldwasser, S., Wigderson, A.: Completeness theorems for non-cryptographic fault-tolerant distributed computation. In: STOC'88, Hong Kong, Chicago, IL, USA, pp. 1–10. ACM (1988)
8. Blum, A., Dwork, C., McSherry, F., Nissim, K.: Practical privacy: the SuLQ framework. In: PODS'05, Baltimore, Maryland, USA, pp. 128–138. ACM (2005)
9. Blum, A., Ligett, K., Roth, A.: A learning theory approach to non-interactive database privacy. In: STOC 08, Victoria, British Columbia, Canada (2008)
10. Boaz Barak, E.A.: Privacy, accuracy, and consistency too: a holistic solution to contingency table release. In: PODS'07, Beijing (2007)
11. Canny, J.: Collaborative filtering with privacy. In: IEEE Symposium on Security and Privacy, San Francisco, Oakland, Ca, USA, pp. 45–57 (2002)
12. Canny, J.: Collaborative filtering with privacy via factor analysis. In: SIGIR'02, Tampere, Tampere, Finland, pp. 238–245. ACM (2002)
13. Chen, H., Cramer, R.: Algebraic geometric secret sharing schemes and secure multi-party computations over small fields. In: CRYPTO 2006, Santa Barbara (2006)
14. Chin, F., Ozsoyoglu, G.: Auditing for secure statistical databases. In: ACM 81: Proceedings of the ACM'81 Conference, Seattle, ACM' 81 is Los Angeles, Ca, USA, pp. 53–59 (1981)
15. Chu, C.T., Kim, S.K., Lin, Y.A., Yu, Y., Bradski, G., Ng, A.Y., Olukotun, K.: Map-reduce for machine learning on multicore. In: NIPS 2006, Vancouver, B.C., Canada (2006)
16. Cohen, W.W.: Enron email dataset. (2004) http://www-2.cs.cmu.edu/~enron/

17. Cohen Benaloh, J.: Secret sharing homomorphisms: keeping shares of a secret secret. In: CRYPTO'86, Santa Barbara, pp. 251–260 (1987)
18. Cormode, G.: Personal privacy vs population privacy: learning to attack anonymization. In: KDD'11, Chicago, pp. 1253–1261. ACM, New York (2011)
19. Cramer, R., Damgård, I.: Zero-knowledge proof for finite field arithmetic, or: can zero-knowledge be for free? In: CRYPTO'98, San Diego. Springer (1998)
20. Dalenius, T.: Towards a methodology for statistical disclosure control. Statistik Tidskrift **15**, 429–444 (1977)
21. Damgård, I., Ishai, Y., Krøigaard, M., Nielsen, J.B., Smith, A.: Scalable multiparty computation with nearly optimal work and resilience. In: CRYPTO 2008, Santa Barbara, pp. 241–261 (2008)
22. Das, A.S., Datar, M., Garg, A., Rajaram, S.: Google news personalization: scalable online collaborative filtering. In: WWW'07, Geneva, Banff, Alberta, Canada, pp. 271–280. ACM (2007)
23. Dhanjani, N.: Amazon's elastic compute cloud [ec2]: initial thoughts on security implications. http://www.dhanjani.com/archives/2008/04/
24. Dinur, I., Nissim, K.: Revealing information while preserving privacy. In: PODS'03, San Diego, San Diego, California, pp. 202–210 (2003)
25. Du, W., Zhan, Z.: Using randomized response techniques for privacy-preserving data mining. In: KDD'03, Washington DC, pp. 505–510. ACM, New York (2003)
26. Du, W., Han, Y., Chen, S.: Privacy-preserving multivariate statistical analysis: linear regression and classification. In: SDM 04, Toronto, Lake Buena Vista, Florida, USA, pp. 222–233 (2004)
27. Duan, Y.: Privacy without noise. In: CIKM'09, Hong Kong. ACM, New York (2009)
28. Duan, Y., Wang, J., Kam, M., Canny, J.: A secure online algorithm for link analysis on weighted graph. In: Proceedings of the Workshop on Link Analysis, Counterterrorism and Security, SDM 05, Newport Beach, pp. 71–81 (2005)
29. Duan, Y., Canny, J.: Zero-knowledge test of vector equivalence and granulation of user data with privacy. In: IEEE GrC 2006, Atlanta (2006)
30. Duan, Y., Canny, J.: Practical private computation and zero-knowledge tools for privacy-preserving distributed data mining. In: SDM'08, Atlanta (2008)
31. Duan, Y., Canny, J.: How to deal with malicious users in privacy-preserving distributed data mining. Stat. Anal. Data Min. **2**(1), 18–33 (2009)
32. Duan, Y., Canny, J., Zhan, J.: P4P: Practical large-scale privacy-preserving distributed computation robust against malicious users. In: USENIX Security Symposium 2010, San Francisco, Washington, D.C, pp. 609–618 (2010)
33. Dwork, C.: An ad omnia approach to defining and achieving private data analysis. In: PinKDD, San Jose, pp. 1–13 (2007)
34. Dwork, C.: Ask a better question, get a better answer a new approach to private data analysis. In: ICDT 2007, Barcelona, Spain, pp. 18–27. Springer (2007)
35. Dwork, C., Kenthapadi, K., McSherry, F., Mironov, I., Naor, M.: Our data, ourselves: privacy via distributed noise generation. In: EUROCRYPT 2006, Saint Petersburg. Springer (2006)
36. Dwork, C., McSherry, F., Nissim, K., Smith, A.: Calibrating noise to sensitivity in private data analysis. In: TCC 2006, New York, pp. 265–284. Springer (2006)
37. Evfimievski, A., Gehrke, J., Srikant, R.: Limiting privacy breaches in privacy preserving data mining. In: PODS'03, San Diego, pp. 211–222 (2003)
38. Feigenbaum, J., Nisan, N., Ramachandran, V., Sami, R., Shenker, S.: Agents' privacy in distributed algorithmic mechanisms. In: Workshop on Economics and Information Securit, Berkeley (2002)
39. Fiat, A., Shamir, A.: How to prove yourself: practical solutions to identification and signature problems. In: CRYPTO 86, Santa Barbara, California, USA (1987)
40. Fitzi, M., Hirt, M., Maurer, U.: General adversaries in unconditional multi-party computation. In: ASIACRYPT'99, Singapore (1999)
41. Ganta, S.R., Kasiviswanathan, S.P., Smith, A.: Composition attacks and auxiliary information in data privacy. In: KDD'08, Las Vegas, pp. 265–273. ACM, New York (2008)

42. Goldreich, O.: Foundations of Cryptography: Volume 2 – Basic Applications. Cambridge University Press, Cambridge (2004)
43. Goldreich, O., Micali, S., Wigderson, A.: How to play any mental game. In: STOC'87, New York, pp. 218–229 (1987)
44. Goldreich, O., Oren, Y.: Definitions and properties of zero-knowledge proof systems. J. Cryptol. 7(1), 1–32 (1994)
45. Goldwasser, S., Micali, S., Rackoff, C.: The knowledge complexity of interactive proof systems. SIAM J. Comput. 18(1), 186–208 (1989)
46. Goldwasser, S., Levin, L.: Fair computation of general functions in presence of immoral majority. In: CRYPTO'90, Santa Barbara, pp. 77–93. Springer (1991)
47. Hirt, M., Maurer, U.: Complete characterization of adversaries tolerable in secure multi-party computation (extended abstract). In: PODC'97, Santa Barbara (1997)
48. Hirt, M., Maurer, U.: Player simulation and general adversary structures in perfect multiparty computation. J. Cryptol. 13(1), 31–60 (2000)
49. Kargupta, H., Datta, S., Wang, Q., Sivakumar, K.: On the privacy preserving properties of random data perturbation techniques. In: ICDM'03, Melbourne, Florida, USA, p. 99. IEEE Computer Society, Washington (2003)
50. Kearns, M.: Efficient noise-tolerant learning from statistical queries. In: STOC'93, San Diego, pp. 392–401 (1993)
51. Kifer, D., Machanavajjhala, A.: No free lunch in data privacy. In: SIGMOD'11, Athens, Greece, pp. 193–204. ACM, New York (2011)
52. Kleinberg, J., Papadimitriou, C., Raghavan, P.: Auditing boolean attributes. In: PODS'00, Dallas, pp. 86–91. ACM, New York (2000). doi:http://doi.acm.org/10.1145/335168.335210
53. Lehoucq, R.B., Sorensen, D.C., Yang, C.: ARPACK users' guide: solution of large-scale eigenvalue problems with implicitly restarted Arnoldi methods. SIAM, San Francisco (1998)
54. Li, N., Li, T., Venkatasubramanian, S.: t-closeness: privacy beyond k-anonymity and l-diversity. In: Proceedings of the IEEE 23rd International Conference on Data Engineering, Istanbul, pp. 106–115 (2007)
55. Lindell, Y., Pinkas, B.: Privacy preserving data mining. J. Cryptol. 15(3), 177–206 (2002)
56. Lindell, Y., Pinkas, B., Smart, N.P.: Implementing two-party computation efficiently with security against malicious adversaries. In: SCN'08, Amalfi, Italy (2008)
57. Liu, W.M., Wang, L.: Privacy streamliner: a two-stage approach to improving algorithm efficiency. In: CODASPY'12, San Antonio, pp. 193–204. ACM, New York (2012)
58. Machanavajjhala, A., Kifer, D., Gehrke, J., Venkitasubramaniam, M.: l-diversity: privacy beyond k-anonymity. In: Proceedings of the IEEE 22rd International Conference on Data Engineering, Atlanta (2006)
59. Malkhi, D., Nisan, N., Pinkas, B., Sella, Y.: Fairplay—a secure two-party computation system. In: SSYM'04: Proceedings of the 13th Conference on USENIX Security Symposium, San Diego, CA, pp. 20–20. USENIX Association, Berkeley (2004)
60. McSherry, F.: Privacy integrated queries: an extensible platform for privacy-preserving data analysis. Commun. ACM 53(9), 89–97 (2010)
61. McSherry, F., Mironov, I.: Differentially private recommender systems: building privacy into the netflix prize contenders. In: KDD'09, Paris, pp. 627–636 (2009)
62. McSherry, F., Talwar, K.: Mechanism design via differential privacy. In: FOCS'07 Rhode Island (2007)
63. Nergiz, M.E., Atzori, M., Clifton, C.: Hiding the presence of individuals from shared databases. In: SIGMOD'07, Beijing, pp. 665–676. ACM, New York (2007)
64. Nissim, K., Raskhodnikova, S., Smith, A.: Smooth sensitivity and sampling in private data analysis. In: STOC'07, El Paso, Texas, USA, pp. 75–84. ACM (2007)
65. Paillier, P.: Trapdooring discrete logarithms on elliptic curves over rings. In: ASIACRYPT'00, Kyoto (2000)
66. Pedersen, T.: Non-interactive and information-theoretic secure verifiable secret sharing. In: CRYPTO'91, Santa Barbara (1992)

67. Pinkas, B., Schneider, T., Smart, N., Williams, S.: Secure two-party computation is practical. Cryptology ePrint Archive, Report 2009/314 (2009)
68. Samarati, P., Sweeney, L.: Generalizing data to provide anonymity when disclosing information (abstract). In: Proceedings of the Seventeenth ACM SIGACT-SIGMOD-SIGART Symposium on Principles of database systems, PODS'98, Seattle, p. 188. ACM, New York (1998). doi:10.1145/275487.275508. http://doi.acm.org/10.1145/275487.275508
69. Samarati, P., Sweeney, L.: Protecting privacy when disclosing information: k-anonymity and its enforcement through generalization and suppression. Technical Report SRI-CSL-98-04, SRI International (1998)
70. Stewart, G.W., Sun, J.G.: Matrix Perturbation Theory. Academic, Boston New York (1990)
71. Sweeney, L.: k-anonymity: a model for protecting privacy. Int. J. Uncertain. Fuzziness Knowl.-Based Syst. $10(5)$, 557–570 (2002)
72. Trefethen, L.N., III, D.B.: Numerical Linear Algebra. SIAM, Philadelphia (1997)
73. Vaidya, J., Clifton, C.: Privacy-preserving k-means clustering over vertically partitioned data. In: KDD'03, Washington DC (2003)
74. Wright, R., Yang, Z.: Privacy-preserving bayesian network structure computation on distributed heterogeneous data. In: KDD'04, New York, pp. 713–718 (2004)
75. Xiao, X., Tao, Y.: M-invariance: Towards privacy preserving re-publication of dynamic datasets. In: SIGMOD 2007, Beijing, pp. 689–700 (2007)
76. Yang, Z., Zhong, S., Wright, R.N.: Privacy-preserving classification of customer data without loss of accuracy. In: SDM 2005, Newport Beach (2005)
77. Yao, A.C.C.: Protocols for secure computations. In: FOCS'82, Chicago, pp. 160–164. IEEE (1982)
78. Zhang, L., Jajodia, S., Brodsky, A.: Information disclosure under realistic assumptions: privacy versus optimality. In: CCS'07, Alexandria, pp. 573–583 (2007)

Index

A. Gkoulalas-Divanis and A. Labbi (eds.), *Large-Scale Data Analytics*,
DOI 10.1007/978-1-4614-9242-9, © Springer Science+Business Media New York 2014

Printed in the United States
By Bookmasters